SOCIOLOGICAL CHALLENGES
n. 1

Editor-in-chief
Liana M. Daher, University of Catania (Italy)

Members of the Scientific Committee
Francesco Antonelli, University Roma Tre (Italy), Kaan Agartan, Framingham State University (USA), Rita Bichi, Catholic University of the Sacred Heart, Milan (Italy), Tova Benski College of Management Academic Studies, Tel Aviv (Israel), Giuseppina Cersosimo, University of Salerno (Italy), Vincenzo Cicchelli, University Paris-Descartes (France), Augusto Gamuzza, University of Catania (Italy), Gennaro Iorio, University of Salerno (Italy), Anna Maria Leonora, University of Catania (Italy), Fabio Lo Verde, University of Palermo (Italy), Giorgia Mavica, University of Catania (Italy), Kevin McDonald, Middlesex University (United Kingdom), Santina Musolino, University Roma Tre (Italy), Sergio Severino, Kore University of Enna (Italy), Geoffrey Pleyers, Catholic University of Louvain (Belgium), Camilo Tamayo Gómez, University of Leeds (United Kingdom), Ligia Tavera Fenollosa, FLACSO - Facultad Latinoamericana de Ciencias Sociales (Mexico), Benjamin Tejerina Montaña, University of Basque Countries (Spain)

UNDERSTANDING SOCIAL CONFLICT
The Relationship between Sociology and History

Edited by
Liana M. Daher

© 2020 – MIMESIS INTERNATIONAL
www.mimesisinternational.com
e-mail: info@mimesisinternational.com

Book series: *Sociological Challenges*, n. 1

Isbn: 9788869771613

© MIM Edizioni Srl
P.I. C.F. 02419370030

TABLE OF CONTENTS

INTRODUCTION
THE RELATIONSHIP BETWEEN SOCIOLOGY AND HISTORY
IN UNDERSTANDING CONFLICT EVENTS   9
Liana M. Daher

METHODOLOGY AND THEORY ISSUES

1. BIOGRAPHICAL FIELD AND THE MEMORY OF THE PAST   25
Rita Bichi

2. THE CONSEQUENCES OF SOCIAL MOVEMENTS.
METHODOLOGICAL LINKS BETWEEN SOCIOLOGY AND HISTORY   33
Liana M. Daher

3. THE CONSTRUCTION OF SOCIAL ACTION.
FROM THE POSITIVISTIC TO THE VOLUNTARIST THEORY OF THE ACTION   47
Paolo De Nardis

4. THEORETICAL THOUGHTS AND HISTORICAL CONTEXT:
THE CASE OF SOCIAL RELATIONSHIPS   57
Gennaro Iorio

5. THE RELATIONSHIP BETWEEN SOCIOLOGY
AND HISTORY IN THE CLASSICS OF SOCIOLOGY   73
Giorgia Mavica, Davide Nicolosi and Alessandra Scieri

SOCIAL AMBIVALENCES AND CONFLICTS IN EMPIRICAL SOCIO-HISTORICAL RESEARCH

AMBIVALENCES AND CONFLICTS IN SOCIO-HISTORICAL PROCESS

6. "CHANGE BEGINS WHERE SILENCE IS BROKEN".
DYNAMICS OF CONTINUITY AND CHANGE IN THE
WOMEN'S PEACE MOVEMENT IN ISRAEL 1983- 2010   105
Tova Benski

7. POPULAR POPULISM   129
Antimo Luigi Farro and Simone Maddanu

8. A "NEW MANIA". THE CARBONERIA:
CONFLICTS AND AMBIVALENCES (SICILY, 1820-1830)   141
Elena Frasca

9. CIVIL SOCIETY VERSUS INSTITUTIONS:
THE HISTORICAL-SOCIOLOGICAL UNDERSTANDING OF A SOCIAL
MOVEMENT BASED ON NON-VIOLENT EXTRA-INSTITUTIONAL TACTICS   155
Sergio Severino and Giada Cascino

10. NAPLES AND THE REVOLUTIONS OF 1799: A CLARIFICATION
OF THE FACTS BY TRUTHFUL ACCOUNTS OF SOME EYEWITNESSES   171
Cinzia Recca

11. A WAR BASED ON SLANDER, TRICKS AND QUIBBLES. THE CONFLICT
BETWEEN HOMEOPATHY AND ALLOPATHY IN THE BOURBON SOUTH   193
Silvana Raffaele

12. A COMPARISON OF ARCHIVE DOCUMENTS: POWER CONFLICTS
REGARDING THE ESTABLISHMENT OF NEW BISHOPRICS IN SICILY
(18TH-19TH CENTURIES)   209
Mariaelena Costa

COSMOPOLITANISM AND SOCIAL CONFLICTS

13. FROM GLOBALIZATION OF CULTURE TO AESTHETICO-CULTURAL
COSMOPOLITANISM   223
Vincenzo Cicchelli and Sylvie Octobre

14. COSMOPOLITAN SOLIDARITY PRACTICES AND IDENTITY CONFLICTS.
THE CASE OF INTERNATIONAL VOLUNTEERS FOR DEVELOPMENT
IN TANZANIA AND MADAGASCAR   241
Augusto Gamuzza

CONFLICT IN EDUCATION

15. THE ITALIAN FAMILY: HOUSEHOLDS, RELATIONSHIPS AND CONFLICTS
(XVIITH-XIXTH CENTURIES)   261
Giovanna Da Molin and Angela Carbone

16. INTERDISCIPLINARY CLASSROOMS AS INNOVATIVE SPACES
TO RESEARCH AND DISCUSS CONFLICTS AND POSSIBLE CHANGES   281
Giovanna Summerfield

17. SOFT REVOLUTION! EDUCATION SYSTEMS IN CONFLICT: STANDARD
VERSUS NON-STANDARD FORMS OF SOCIALIZATION IN A GENDER PERSPECTIVE   291
Anna Maria Leonora

AUTHORS BIOGRAPHICAL PROFILES   311

Liana M. Daher

Introduction
The Relationship Between Sociology and History in Understanding Conflict Events

The indissoluble link between sociology and history is not new to the social sciences. In the early 20th century, Weber (1922) saw relationships between history and sociology as based on mutual and essential support and logical priority, according to which, rephrasing Alessandro Cavalli (2001), "sociology without history is blind, history without sociology is mute."

In fact, following the intentions of its Founding Fathers, the discipline of sociology arose as a "science of connections" aiming at investigating relationships between social life phenomena and events, even those apparently far from each other. According to a strategic interdisciplinary analysis, also the "Annales" lesson confirmed the indissoluble link between history and social sciences. Braudel was a great defender of the view that history and sociology are a single unitary enterprise: «one single intellectual adventure, not two different sides of the same cloth but the very stuff of that cloth itself, the entire substance of its yarn» (Braudel 1980: 69).

Conversely, in his famous book *Sociology and History. Controversies in sociology* (1980), Burke gives a very synthetic definition of this relationship:

> Sociology may be defined as the study of human society, with an emphasis on generalizations about its structure and development. History is better defined as the study of human societies in the plural, placing the emphasis on the differences between them and also the changes which have taken place in each one over time (Burke 1980: 2).

He believes in a dangerous narrowing of perspective in which historians tend to perceive the problem as something unique, and not a combination of elements with correspondences in other places and times, while sociologists tend to generalize everything through the lens of contemporary experience, not paying attention to the perspective of long-term historical processes and social changes. He perceives two different and incompatible positions where sociologists attribute importance to the numbers, recognizing the rules of variations, and historians give priority to the words, stressing the individual and specific (Ibidem: 9-11). He looks at the possible relationships of sociology and history in a very critical way, highlighting a certain resistance to mutual cooperation, even though their mutual interrelations should be fruitful and advantageous to both.

A clear-cut but simplistic answer to the relationship between the two approaches is that: "Sociology is nomothetic, while history is idiographic", i.e., the historian describes unique events, while the sociologist derives generalizations from the social world[1]. While recognizing the above synthetic definition as quite true, the challenge of this book is to go ahead and overcome the mere separation between nomothetic and ideographic knowledge. The differences and similarities of the two disciplines will be investigated in a dialectic way in order to find synthesis and fruitful connections.

History uses generalizations to examine particular sequences of events and, like sociology, often analyzes events through a biographical and hermeneutical lens. Therefore, also regarding the method, historiography and sociology cannot be radically separated: they analyze the same subject matter sometimes from the same point of view and through ways that the two continue to borrow from each other extensively. Works like Weber's *Protestant Ethic and Spirit of Capitalism*, and Sorokin's *Social and Cultural Dynamics* show that the line for demarcation between history and sociology can only be blurred.

The analysis of the theoretical and methodological differences and similarities of the two disciplines calls for care and method. The scientific

---

[1] The debate regarding the separation between history and sociology had its origin in classical thought. Even Durkheim (1888), stressing each domain difference based on the fact that historians cannot have a general approach and cannot always formulate laws, recognizes the importance of the scientific work of history as a base to respond to sociological issues, starting from the historical hypothesis of connections and causes in past events. Marx and Weber took human history as their field of inquiry, trying to answer such grand questions as "what are the historical or evolutionary sources of modern society?" (Smelser, Warner 1976: 9). Marx rooted his theory of society in historical materialism and conflict, while Weber (1922) made use of a comparative method based on the ideal type model, on which the solidarity between sociology and history can be based, that can still be applied to the contemporary study of society. To obtain more in-depth explanations about the views of the Founding Fathers of sociology regarding the connections between sociology and history, please see chapter five of this book.

literature is very extensive in this sense[2]. Both sociology and modern historiography had their origin in the 19th century, respectively establishing the concept of historical periods and the notion of historical types of society. The interaction between the two disciplines can be found in their subject matters which are interdependent and overlap to a considerable extent.

On the one hand, historians frequently provide the material employed by sociologists – historical sociology depends upon the data that only a historian can supply, and the comparative method often requires historical data –; on the other hand, sociological research provides historians with several kinds of information, and the subject matter of social history overlaps to a very great extent with sociology, in particular historical sociology.

To summarize: the two disciplines differ in the way of looking at social events, aiming at complementary scientific results. History is mainly concerned with past events, aiming at systematically recording the story of humanity. Past human society and social processes represent their principal objects of study, dealing with past events and studying past social, political, and economic aspects. The investigation of present and complex social phenomena is instead the main interest of sociology. It is concerned with the study of the historical development of societies, investigating various stages of life, lifestyles, traditions, habits, etc., and their organization in social institutions.

According to a widely accepted point of view, a self-respecting sociologist should deal with the present and not with the past. This can be investigated by focusing on the "here and now", unlike the historian whose focus is the "how it was", or in a long-term perspective, as will be demonstrated by some of the contributions in this book.

As Weber (1922a) rightly pointed out, causality lies at the core of the relationship between sociology and history:

> The sociology of happening needs the historian's sense of the complex but finally casual phasing of action. The history happening needs the sociology's sense of the remote but cogent casual weight of structure. Both need an overt, simple and self-conscious capacity both to represent and account for significant sequence which neither as yet fully possesses (Abrams 1982: 315).

Historical and sociological causality are closely connected, both expressing themselves in terms of possibility. The probability issue could

---

[2] Just to recall some challenging contributions to the debate, see Wolff 1959; Means 1962; Prandstraller 1969; Restivo 1970; Crespi 1974; Giddens 1979; Abrams 1982; Skocpol 1985; Cavalli 1989; Goldthorpe 1991, 1994.

be considered a mediator factor between the two disciplines, and between the *freedom* and *necessity* approaches. In fact, the historian usually refers to a number of determining factors in explaining why a specific course of action was chosen among several alternatives; the sociologist adds additional considerations to this analysis based on the model of rational social action, as developed by Weber (1922b), and the social constraints coming from human interaction. In this sense, the actor's choice will be both free and understandable, excluding the *fate* factors from the analysis (Cahnman and Boskoff 1964: 5-6, emphasis added by the author).

Causality and probability issues open the discourse on the common aim of studying social change. History *is* change, and social change theories go in the direction of identifying socio-historical factors to explain/understand specific courses of events (De Nardis 1998: 57, emphasis added by the author). History aims at rebuilding social events in evolutionary processes, while sociology arose as the science that mainly studies social change. The need has emerged to rebuild social processes in a longitudinal direction, through research methods, data, and interpretations characterized by a dynamic relationship between temporal registers: an analysis where it is possible to distinguish between past, present, and future as fundamental and inseparable parts of the temporal flows, but at the same time establishing a narrative continuity among these three temporal spaces in order to understand human action.

Time becomes a crucial and founding element of the relationship between sociology and history. Time, and the perception of it, marks the relationship between the two disciplines, and the two temporally different ways of investigating social reality, albeit in a continuity relationship.

As already highlighted by Park and Burgess:

> [History] seeks to reproduce and interpret concrete events as they actually occurred in time and space. [Sociology] seeks to arrive at natural laws and generalizations in regard to human nature and society, irrespective of time and place. [History] seeks to find out what actually happened and how it all came about. [Sociology] seeks to explain, on the basis of a study of other instances, the nature of the process involved (1921: 11).

The idea of process is crucial in order to achieve successful results in sociological work. Facing historical changes, it shaped understanding through reading the experience of those changes (Abrams 1982: 3). The basic assumption is that change is the main feature of human societies. Thus, «Laws of history [...] must be laws of change» (Bock 1964: 23), concerning both historical and sociological knowledge.

We cannot deny that the cognitive interest of historians and sociologists considers reality from a different perspective that mainly concerns the different meanings given to the nature and role of conceptual and theoretical categories. This difference acquires a different meaning at a methodological level, and the difference – and integration – between quantitative and qualitative methods is played on (Cavalli 1989). A possible convergence can be found, as proposed in Chapters 1 and 2, in the biographical approach, that breaks down certain types of barriers artificially erected between the two disciplines (Macioti 1989: 249). Needless to say, making use of the comparative method, following Weber's theory, represents one of the most fruitful encounters between sociology and history.

Borrowing the *bricolage* metaphor from Simmel's theory, it is possible to highlight some connections between the method of sociology and that of history. According to Weinstein & Weinstein (1991: 164), we can see *bricolage* as a complex, dense, reflexive, collage-like creation that represents the researcher's understandings and interpretations of the phenomenon/event under analysis. The role of sociology and history researchers is to connect the parts to the whole, to stress the meaningful relationships in the social processes investigated. In addition, it should be highlighted that the historical use of accounts is often something more than the perspective of the individual observer (Luhmann 1986). According to Tuchman (1994: 306), the presentation of ideal types historically and sociologically situated «includes an interpretative framework that implicitly contains some notion of the 'meaning of history'».

As stated by Mills (1959: 145), «Every social science – or better, every well considered social study – requires an historical scope of conception and a full use of historical materials»; in this sense, the understanding of a social event requires the use of general propositions that have to be demonstrated starting from historical analysis and comparisons (Busino 1975: 63).

Thus, our challenge concerns the quest for elements of interaction between the disciplines and, at the same time, useful compromises in the interdisciplinary study of phenomena and common fields of research. Taking into account Bourdieu's remark, according to which «The separation between sociology and history is devastating and totally devoid of epistemological justification: every sociology must be historical and every history must be sociological» (Bourdieu and Loïc 1992: 62), we should consider that this does not mean that history and sociology are the same thing (Abrams 1982: X) or that there are no logical or even methodological distinctions between the social sciences and history (Giddens 1979: 230).

It means instead that history can learn from sociology, and vice versa. As we will see through the different contributions in this book, they work on a common ground of research – the human society – sometimes making use of similar approaches and similar goals, but remaining «two significantly different intellectual enterprises» (Goldthorpe 1991: 225). We should not see them as completely separate disciplines, but try to underline the compromises and differences through concrete cases of empirical research, that is the principal aim of this book.

Social conflict, designed also as social ambiguities and/or ambivalences, will be the subject terrain on which this exchange will be held. Social conflict is a common ground of research both for sociology and history, through which the book aims at providing argumentative issues to the challenge represented by the relationship between history and sociology, and shows meaningful convergences between the two disciplines in order to offer innovative spaces of discourse around the theory and methodology of research, on the one hand, and some areas of yesterday's and today's social conflicts, on the other. Social conflict is a common topic and represents a conceptual framework of issues such as: collective action, protest, social exclusion, cooperation, etc., on which this book focuses.

Looking at human sciences studies on conflict, a key question emerges: is the conflict a positive or a negative (i.e. destructive) force? We cannot avoid recalling Simmel's concept of ambivalence, both historical and sociological, but one cannot also forget how the idea of conflict as a principle of current positive and driving force goes back to the origins of philosophical thought.

As Heraclitus argues: «We must recognize that war is common, strife is justice, and all things happen according to strife and necessity» (58, b80). Going back to Simmel, regarding conflict, two parallel and distinct tendencies of human beings should be outlined: the associative one and the dissociative one, which are more easily separated in abstract analysis than in empirical facts (Turner 1994: 145). Both tendencies concern conflict: conflicting actions are interactions between individuals, and conflict can play an integrative function well (Simmel 1908, Eng. tr. 1956: 23-24).

The above perspective underlines the category of ambivalence as a conceptual pivot useful to effectively achieve an analytical interpretation of conflict. Social phenomena are intrinsically ambiguous and contradictory, and this category can read them both well through the sociological and historical lens; this is even more true in the analysis of explicit conflict events.

The combination of ambivalence and conflict has played a key role since the Hegelian dialectic turn (1812-16), followed by the socio-economic contribution of Marx (1867-1894), and, as stressed above, in the sociological theory of Simmel (1908), where the "marriage" between ambivalence and conflict takes shape as a category within relational contexts denoted as *sociation*. After finding vital application in the *Sociological Ambivalence* (1976) of Merton, this heritage – concerning theory and scientific knowledge – seems now to have fallen out of use (Levine 1985; Calabrò 1997), at least in the sociological approaches.

The proposal to analyze the different areas in which ambivalence can be used – both as a heuristic interpretation key (*frame*) and means of connection (*bridge*) and to explain social interactions – comes from the assumption that ambivalence is one of the "reflexive hearts" of the human sciences, in particular sociology and history. The *fil rouge* is that by making use of their respective competences, it would be possible to achieve most effective and comprehensive results, even though the two disciplines share different objects of investigation.

The book aims at developing this cooperative field by dealing with two big common research sections: Methodology and Theory Issues, and Social Ambivalences and Conflicts in empirical socio-historical research. The latter is divided into three sub-sections: Ambivalences and conflict in historical process, Cosmopolitanism and Social Conflicts, and Conflict in Education.

The Methodology and Theory Issues section includes several papers addressing the debate on the relationship between sociology and history with regard to several aspects of the fieldwork of human sciences researchers. The topics dealt with are both related to the above relationship and the subjects of ambivalence and conflict, the analyzing methods, and particular perspectives related to the interconnection of the two research areas.

The decision to open the book with the methodological papers has been outlined above. Convergences and meaningful exchange will be possible in sharing a biographical approach even if with different aims. The first two chapters (Bichi, Daher) address their argumentation exactly in this direction; the first presenting the expression "biographical field" as adequate to find homogeneity in the different styles of putting in practice the method, stressing the longitudinal dimension declined at an individual and social level; the second proposing a special combination of biographical methods taking advantage of the sociology-history link in studying social movements as longitudinal processes of conflict, mixing two different temporal lenses of observation. Following this methodological beginning, the theoretical chapters propose different readings of the relationship between sociology and history. De Nardis identifies in Labriola's works the theoretical innovation able to overcome the separation between nomothetic and ideographic sciences, and the explanatory key to the relationship between sociology and history. Iorio aims to highlight the fact that "social relationships" have been at the basis of sociological study ever since sociology was established as a science of human behavior, and uses this evidence to create a boundary which distinguishes it from philosophy, law, psychology, biology, economics, history, and politics, all of which concern the interpretation of social phenomena. The chapter by Mavica, Nicolosi and Scieri closes the first section, focusing on and analyzing the connections between sociology and history as developed by three of the Founding Fathers of sociology: Marx, Simmel, and Weber, considered among the most significant scholars regarding both the relationship between sociology and history and the subject of conflict.

The *Social Ambivalences and Conflicts in empirical socio-historical research* section welcomes several fieldwork proposals, investigating conflict in sociological and historical research, making use, and sometimes interrelating, the assumptions and methods of both approaches. The reader can find substantial evidence of the debate in this section.

In the three sub-sections of the book, the research works of sociologists and historians follow each other, creating a constellation of proposals. In the first chapter of the first sub-section – Ambivalences and conflict in historical process – Benski provides substantial evidence of the Women's Peace Movement in Israel from 1983 to 2010 through an almost lifelong study, taking advantage of the perspective of triangulation of methods and activists, producing a real exchange between the historical and sociological approaches. Farro and Maddanu attempt to identify a specific form of populism as a historical, cultural, and social phenomenon, that has been appropriated by political actors who have successfully interpreted it to lead a general agenda, while considering specific contemporary and conflicting topics that have arisen in the age of globalization. Through first-hand sources, Frasca analyzes the Carboneria riots of 1820-21 in Southern Italy as an interesting key to understand how clear and unclear conflicts made for socio-political mutation in those tumultuous and important years. On the sociological front, Severino and Cascino propose an historical-sociological understanding of Danilo Dolci's movement that took place through non-violent non-institutional tactics in Sicily (Valle del Belìce) in the period between the 1950s and 1970s, highlighting the mutual relationship between biographies and historical events in the social structure. Still on the protests theme, Recca analyzes several episodes of social and political unrest in Naples (1799) including a revolution involving large sectors of the population and the counter-revolution that followed it. The account is based on how the historical interpretations of these catastrophic struggles was experienced and recorded by important figures, men and women belonging to different social classes. The meaning of the concept of conflict changes in the last two chapters of the sub-section (Raffaele and Costa), taking the sense of ambivalence. The former examines the struggle between homeopathy and allopathy during the Enlightenment in medicine and politics, examining complaints, conflicts, and new theories in a renewed holistic view that looks at sick people *in toto*; the latter focuses on the establishment of new bishoprics in Sicily during the 18th and 19th centuries as an example of State and urban power conflicts and debates comparing Church archive documents to find new methods, perspectives, and historiographical views.

The sub-section Cosmopolitanism and Social Conflicts consists of two chapters, focusing on the longitudinal cosmopolitan perspective in social sciences. Cicchelli and Octobre propose a specific way to analyze the effect of the globalization of culture on the way young people see the world, by examining their seemingly banal consumption for clues to understand how they envision the world, and by introducing a specific approach stemming from cosmopolitanism and the sociology of culture: the aesthetico-cultural amateurship. Gamuzza focuses instead on the relevance of a *cosmopolitan solidarity* to global issues and social change, stressing that when this form of solidarity is translated into biographies and historic/personal trajectories, it implies a conflicting outcome upon the subject's identity.

The last sub-section, Conflict in Education, ends the book with three interesting chapters. Da Molin and Carbone aim to reconstruct some demographic and social aspects of the Italian family in the Modern Age, focusing on households in different economic, geographical, and social contexts, and on the relationships and conflicts within the family. Summerfield shows, through the results of practical activities used in an advanced undergraduate course titled "East Meets West: Sicily, a Case Study at the Crossroads" how history matters and interchanges with other human sciences, including sociology, in understanding specific cultures and habits, and the concepts of citizenship and global citizenship, cosmopolitanism, transnationalism, and multiculturalism, and so on. Last but not least, Leonora proposes an exploratory fieldwork regarding education practices that provide an alternative to state schools in Italy, establishing two levels of issues: 1) an up-to-date description of the symbolic elements that surround parental values on the basis of homeschooling and other alternative education practices, and 2) an original interpretation of the theoretical conceptualization of the homeschooling phenomenology. The two issues are analysed as embedded in an historical longitudinal dimension.

The overview of all the proposals of the chapters recalls all the issues regarding the relationship between sociology and history in studying conflict underlined above. All the contributions take into account the questions posed by the research project titled *Ambivalence and Conflict in Belonging and Sociability Spaces from the Modern to the Actual*[3] that gave rise to this book and go beyond trying to find spaces of dialogue and mutual understanding between sociology and history.

A critical reading of this book cannot avoid focusing on the *complementarity* of sociological and historical research because both highlight two fundamental human historicity aspects. History aims at grasping and reconstructing events, while sociology aims at capturing relative social conformities (Abbagnano 1974: 115). As argued before, the arguments on the strict separation between nomothetic and ideographic are not convincing. We can only understand the present by *making it historical*, with the aim of understanding the diachronic dimension in which each social event is embedded; it is the dimension that gives sense to the event. Meanwhile, this dimension can be understood only through proper explanatory constructs and methodological apparatus, i.e., through building a nomological knowledge *whose referent is a more or less long section of diachronicity* (Leonardi 1974: 99). Indeed, the challenge of this book is to give ample exemplification of the above dimension and possible angles of analysis.

Bibliographical references

Abbagnano, N., 'Sociologia, scienza e storiografia', in P. Crespi (ed.), *Sociologia e Storia* (Milano: Celuc, 1974), pp. 103-115.

Abrams, P., *Historical Sociology* (New York: Cornel University Press, 1982).

---

3   The research project was supported by the University of Catania (Program FIRD 2014) and it was concluded by the Conference *Understanding Social Conflict. The Relationship between Sociology and History* held at the Department of Education in December 2016. This book represents in part the output of the research project and Conference. As Principal coordinator of the research and program coordinator of the conference, I want to say thank to all the researchers that have actively and fruitfully cooperated in the project: Silvana Raffaele, Elena Frasca, Augusto Gamuzza and Anna Maria Leonora.

Burke, P., *Sociology and History. Controversies in sociology* (London: George Allen & Unwin, 1980).
Braudel, F., *On History* (London: Weidenfeld & Nicolson, 1980).
Bock, K. E., 'Theories of Progress and Evolution', in W. J. Cahnman and A. Boskoff, *Sociology and History. Theory and Research* (New York: The Free Press of Glencoe, 1964), pp. 21-41.
Bourdieu, P., and L. J. D. Wacquant, *Réponses: Pour une anthropologie réflexive* (Paris: Seuil, 1992).
Busino, G., *Sociologia e storia. Elementi per un dibattito* (Napoli: Guida, 1975).
Cahnman, W. J., and A. Boskoff, 'Sociology and History: Reunion and Rapprochement', in W. J. Cahnman and A. Boskoff, *Sociology and History. Theory and Research* (New York: The Free Press of Glencoe, 1964), pp. 1-18.
Calabrò, A. R., *L'ambivalenza come risorsa* (Bari: LaTerza, 1997).
Cavalli, A., 'Storia e sociologia un incontro difficile', *Studi di sociologia*, 27 (2) (1989), 217-221.
— *Incontro con la sociologia* (Bologna: il Mulino, 2001).
Crespi, P. (ed.), *Sociologia e Storia* (Milano: Celuc, 1974).
De Nardis, P., 'Una nuova rappresentazione per la sociologia scientifica', in P. De Nardis (ed.), *Le nuove frontiere della sociologia* (Roma: Carocci, 1998), pp. 19-77.
Durkheim, É., 'Cours de science sociale: leçon d'ouverture', *Revue international de l'enseignement*, 15 (1888), 23-48.
Giddens, A., *Central Problems in Social Theory* (London: Macmillan, 1979).
Goldthorpe, J. H., 'The Uses of History in Sociology: Reflections on Some Recent Tendencies', *The British Journal of Sociology*, 42 (1991), 211-230.
Goldthorpe, J. H., 'The Use of History in Sociology: A Reply', *The British Journal of Sociology*, 45 (1994), 55-77.
Hegel, G. W. F., *Wissenschaft der Logik* (Nürnberg: Johann Leonhard Schrag, 1812-1816).
Leonardi, F., 'La sociologia', in P. Crespi (ed.), *Sociologia e Storia* (Milano: Celuc, 1974).
Levine, D. N., *The Flight From Ambiguity: Essays in Social and Cultural Theory* (Chicago: University of Chicago Press, 1985).

Luhmann, N., 'The individuality of the individual: Historical meanings and contemporary problems', in T. C. Heller, M. Sosna and D. E. Wellbery (eds), *Reconstructing individualism: Autonomy, individuality, and the self in Western thought* (Stanford, CA: Stanford University Press, 1986), pp. 313-328.

Macioti, M. I., 'Memoria, oralità, vissuto: fra storia e sociologia', *Studi di Sociologia*, 27 (2) (1989), 240-249.

Marx, K., *Das Kapital: Kritik der politischen Oekonomie* (Hamburg: Verlag von Otto Meissner, 1867-1894).

Means, R. L., 'Sociology and History: A New Look at Their Relationships', *The American Journal of Economics and Sociology*, 21 (3) (1962), 285-298.

Merton, R. K., *Sociological Ambivalence and other essays* (New York: The Free Press, 1976).

Mills, C. W., *The Sociological Imagination* (New York: Oxford University Press, 1959).

Park, R., and E. Burgess (eds), *Introduction to the science of sociology* (Chicago: University of Chicago Press, 1921).

Prandstraller, G. P., 'Some convergence in history and sociology', *Social Science Quarterly*, L (1) (1969), 6-58.

Restivo, S. P., 'Sociology And History: Notes On Rapprochement', *The Kansas Journal of Sociology*, 6 (3) (1970), 134-142.

Simmel, G., *Soziologie: Untersuchungen über die Formen der Vergesellschaftung* (Berlin: Dunker & Humblot, 1908).

— *The Streit, in Soziologie: Untersuchungen über die Formen der Vergesellschaftung* (Berlin: Dunker & Humnlot, 1908); Eng. trans. in K. H. Wolff *Conflict and the Web of Group Affiliations* (New York: Free Press, 1956).

Skocpol, T., *Vision and Method in Historical Sociology* (New York and Cambridge: Cambridge University Press, 1984).

Smelser, N. J., and R. S. Warner, *Sociological Theory: Historical and Formal* (N. J.: General Learning Press, 1976).

Sorokin, P. A., *Social and Cultural Dynamics* (New York: American Book Company, 1937-1941).

Tuchman, G., 'Historical Social Science. Methodologies, Methods, and Meanings', in N. K. Denzin and Y. S. Lincoln, *Handbook of Qualitative Research. Thousand Oaks, London* (New Deli: SAGE Publications, 1994), pp. 306-392.

Turner, J. H., 'Marx and Simmel Revisited: Reassessing the Foundation of Conflict Theory', in D. Frisby, *Georg Simmel: Critical Assessments* (Routledge, 1994), vol. III; first ed. in *Social Forces*, 53 (4) (1975), 618-627.

Weber, M., 'Die protestantische Ethik und der "Geist" des Kapitalismus', *Archiv für Sozialwissenschaft und Sozialpolitik*, 20 (1904), S. 1–54 and 21 (1905), S. 1-110; Eng. trans. *Protestant Ethic and Spirit of Capitalism* (New York: Charles Scribner's Sons, 1930).

— *Gesammelte Aufsätze zur Wissenschaftslehre* (Tübingen: Mohr, 1922a); Eng. trans. *On the Methodology of the Social Sciences* (Illinois: The Free Press of Glencoe, 1949).

— *Wirtschaft und Gesellschaft* (Tübingen: Mohr., 1922b); Eng. trans. *Economy and Society: An Outline of Interpretive Sociology* (New York: Bedminster Press, 1968).

Weinstein, D., and M. A. Weinstein, 'Georg Simmel: Sociological Flâneur Bricoleur', *Theory, Culture & Society*, 8 (3) (1991), 151-168.

Wolff, K. H., 'Sociology and History; Theory and Practice', *American Journal of Sociology*, 65 (1) (1959), 32-38.

METHODOLOGY AND THEORY ISSUES

Rita Bichi

1.
Biographical Field and the Memory of the Past

The ways of using biographical data in social research are complex: a heterogeneous and even contradictory mixture, one in which we can detect some opposing issues, a variety of cultural programs, researcher images, and philosophies of research. For these reasons, the expression "biographical field" (Bichi 1999; 2007) seems adequate even to describe its homogeneity in the longitudinal dimension declined at an individual and social level. Time and memory appear to be two of the most important concepts in this area, because the elapsed time and its individual and social memory enter into the reconstructive unfolding of a life as indivisible from the recounted story. With particular attention to the collection of biographical data, the chapter will discuss some hypotheses concerning: the memories involved; the possibility to conceive the past as it is narratively actualized; and individual memory being an appropriate instrument for social research. In Ricoeur (1986) and Bruner's (1995) thought, a narrative makes a product that finds its fullness of meaning in interaction with its user. What are the implications of this interaction when it becomes active in a face-to-face situation?

In Italy, the discussion on social research – that uses few directive tools – is most often defined as qualitative; as early as the 1990s, it began to bear fruit that has since been diversified. The basis of this diversification is the recognition or otherwise of the legitimacy of the stories as material for social research and, ultimately, the degree of trust in the voice of science within the voice of the people. This dichotomy has never been resolved to date, although there have been important attempts to find a third way that includes – in the main quantitative stream still today – narration as a tool, that has started to be legitimate in scientific communities, from which the complex and articulated framework of the so-called mixed methods arise in order to search for a possible integration. This search has yet to be concluded.

Indeed, in the research practice, there are many roads that have been followed in the search for solutions to what has long been considered a problem. One of these is linked – directly or indirectly, consciously or not – to pragmatic approaches that relegate foundational discussions to the background, considering the choice of technique and the tools based only on the purposes and focusing only on the procedures to be followed; another way is that devoted to a radical technicality, which assumes in principle the logical and mathematical language as the only legitimate one because it is considered capable of restoring the reality of social phenomena, relegating any pre-assertive consideration to a 'black box'; a third way is one that substantially indifferent to the questions of the method, adhering to widespread use, and which often produces literature that is even devoid of reports on the research process, especially if it is of a qualitative type; and finally that which starts from theoretical-epistemological approaches legitimizing the story and the experience and which also attempts the difficult systemization of the complex area of biographical techniques and tools available to sociology.

Various models of thought have contributed to the long path of the latter direction, including: Denzin's symbolic interactionism, the Sartriaran Marxism of Ferrarotti, Wallerstein's neo-materialism, the structuralism of Bourdieu and Bertaux, and the hermeneutics of Kohli, without citing the classics from which we always continue to draw inspiration (Campelli 1990). Moreover, in contemporary Western societies, the need for flexible tools, adaptable to highly differentiated contexts, capable of reading complex phenomena in continuous and rapid change, seems to take on evidentiary value. In other words, the typical photographs of the research survey – but also of the most current big data, on which we rely a lot today – are only useful if they are frames of a film, that is, if inserted in a wider repertoire that includes not only the sum of the individualities but also the bringing to light of the articulation between the social and the individual, which, however, requires reconstructed temporal depth and the entry into the field of the process of the production of meanings, obviously from a Weberian point of view.

Biographical material responds to this need but, in international social research, has constituted a complex, uneven and even contradictory whole, in which we can find opposing themes, a multiplicity of cultural programs, images of the researcher, and philosophies of research. For these reasons, the expression "biographical field" seemed appropriate to identify the production and analysis of this material, in which the variety of approaches finds homogeneity only in the temporal-longitudinal dimension, declined precisely at the individual and social levels. I will focus on this last aspect, putting forward some questions rather than providing answers.

Above all, I would say that disciplinary boundaries are of particular importance, because biographies are a way of literary representation of ancient origins; they are important tools of historical analysis, the indispensable basis for identifying exemplary types in psychology, and they constitute the window through which anthropology has looked at unknown worlds. In social research, unlike in literature, biography is not an end in itself, but a vehicle of thought, and unlike psychology, sociology presupposes the existence of a collective referent in which to situate the individual life; for sociological analysis, unlike history, time is not the starting point whose value can be defined *a priori*, but the result of a theoretical choice that brackets the primacy of the present to question itself about how and why the present opens up towards the past that is rebuilt and recognized ex post facto. Despite these – albeit relevant – differences, it is clear that the reflections of the various disciplines are strongly intertwined.

In fact, and in addition to all the differences that mark the biographical field both within sociological thinking and in comparison to other disciplines, it would appear that using biographical material means taking charge of two otherwise flat, hidden, bracketed dimensions: time and the memory of it. Specific to the entire biographical field is what I call *longitudinal intelligibility*: the possibility of reading the associated life longitudinally (Bichi 1999).

The diachronic perspective opens up questions that also concern the direct collection of biographical material, the narration of a person's life, or fragments thereof, to one or more interlocutors. This is the biographical material that sociologists most often use, even if it is not the only one - from Abel's biogram (1947), to Atkinson's life story (1998). Moreover, the forms of memory that are involved are multiple, and their simultaneous presence makes the production situation of the oral story particularly suitable for making them visible. I will try to examine some of the questions I mentioned.

The memory of the past, according to Bertaux (1997), is constantly formed through the patterns of interpretation of the present, and therefore the current stories of past experiences mix the past experience with the experience of the present, and one cannot know, according to this author, if one finds one or the other; it is an approach that resembles that of the early Bourdieu (1986), when he speaks of *illusion biographique*, which supposes the unconscious projection of a present and social-minded representation on the past. Ferrarotti (1997) affirmed that in order to understand and reconstruct the spirit, the atmosphere of a time that no longer exists, to feel in some way truly contemporary with oneself and the people of times gone by, the greatest difficulty is not constituted by what one would need to know, but rather from what one would need to know how to forget, put in brackets, in some way suspend. In the wake of the thought of these authors, we may argue that the field of memory works for losses and continuous acquisitions for overlapping, censorship, and revelation. Are we then able to speak of an objective reconstruction of a fact?

Agostino maintained that in narrating true facts from the past, the reality of past facts is not extracted from memory, but from the words generated by their images, almost footprints impressed by them in our mind as they pass through the senses. Thomas and Znaniecki (1920) argue that the fact itself is already an abstraction when we take it in hand to analyze it; Joutard adds that an event is nothing but the trace it leaves in life and in memories, and that collective memory is woven from these tangled traces. De Gaulejac (1989) maintains that memory forgets, deforms, transforms, and reconstructs what has passed in relation to the need of the unconscious and power and not according to the demands of truth. What kind of past do we know through the narrated memory?

Memory can also be seen as a powerful tool for social integration that does not reproduce or distort a given reality, sub- or conscious; that is, memory is not a mirror, though perhaps even a distorting one, of a real referent. It can be seen rather as a factor in itself of the historical evolution of a society, as symbolic mediation and elaboration of meaning, an act of narration in a social context. In this view, the facts are mediated and time – in its duration – is not a measurable quantity, infinitely divisible, but always takes an associative and emotional coloring, as in the affective memory proper to autobiography (Combe 1987).

So is memory aproblematic – does it require adherence to reality because it is in itself real? Or is it intended as a creator of an artifactual past (Kaha 1989)? Or, yet again, does memory reveal stories that do not occur in the real world, but are constructed through a continuous interpretation and reinterpretation of the experience (Bruner 1995)?

The reconstruction of past experience seems to take place therefore starting from the present, a reflection that is not to be found in the analysis of Thomas and Znaniecki (1920) or Clifford Shaw (1930) or Oscar Lewis (1966). The reason for this is perhaps because these authors (as Bertaux argues) worked with young subjects with a brief history on their shoulders. Does this mean that the age of the subject who remembers influences the remembering processes? That the accumulation of experience changes the field of memory by always generating new ways of remembering?

But in how many ways can the sociological analysis make use of the past time communicated through a story and so variously interpreted? There are at least two large sets of answers.

In the meantime, in a perspective close to structuralism, it is possible to argue that the structure of social relationships, its mechanisms, and the objectivity of relationships that relate groups and classes of individuals can be seen through the narrative. This is the position of Passeron (1990), Bourdieu (1986), Godard (1989) and partly of Bertaux (1997), who despite their differences, consider that the collective biography is not deduced from the individual biographies. The groups (or classes) defined by the objectivity of the relationships – conflict or cooperation, domination or obedience – seem to possess properties and history that are not reduced to those of individuals circulating in groups. Life stories would therefore be pre-structured by organized external temporalities which follow independent causal chains that existed before individual lives, and the narrator identifies the variable form of typical behavior, the support of external processes with their own temporality.

However, is it possible to separate event and experience? Experience and sense of experience? Can we disregard the sense, the identities, the relationships, and the context?

In another perspective, aimed at reconstructing narrative processes and making use of hermeneutics and constructionism, the search for uniqueness and meaning in a past seen from today is interpreted as the necessary overcoming of the original judgment, in how much the truth of experience would always contain the reference to a new experience (following the thought of Gadamer). The reflection on experience allows the subject to relive the original judgment included in the experience recalled and to overcome it in order to experience its validity through time. The relationship between memory, time, and story is famously addressed by Ricoeur, according to whom a circular relationship combines narrativity and temporality; time becomes human as expressed according to a narrative form, and the story reaches its full meaning when it becomes a condition of temporal existence. In other words, as Bruner (1995) states, there is no other way than narrative to describe time lived.

Several questions arise from here: if the story is imitation – a seeming shadow, acceptable mask, continuous creation, or mimesis – and if it is in the narrative that we trace the image of identity, nothing of the self, except in the present story, is it both firm and stable? Or can we think of constructing the image of denser clumps of sediment over time that oppose greater resistance to reflective waves and do not allow them to penetrate if not when they acquire unusual and prolonged power?

In any case, science cannot be limited to "the long silences", those avoided by Popper in Vienna, and I seem to be able to conclude with a common invitation to the exercise of de-localization of the gaze, of being caught not only by cognitive dissonances that open spaces for reflection, not so much from the much-vaunted moments of serendipity, not only with the massive use of abduction but from what can be seen by putting oneself "in another's shoes" (Bourdieu 1993), an operation necessary to understand everyone, in the awareness of at least the partiality of the look of science.

Bibliographical references

Abel, T., 'The nature and use of biograms', *American Journal of Sociology*, LIII (1947).
Atkinson, R., *The life story interview* (London: Sage, 1998).
Bertaux, D., *Rècits de vie* (Paris: Nathan, 1997).
Bichi, R., 'Campo biografico e intelligibilità longitudinale', *Studi di Sociologia*, XXXVII (January-March 1999), 27-54.
— *L'intervista nella ricerca sociale* (Milano: V&P, 2007).
Bourdieu, P., 'L'illusion biographique', *Acte de la recherche en sciences sociales*, 62-63 (1986), 69-72.
Bourdieu, P., *La misère du monde* (Paris: Seuil, 1993).
Bruner, J. S., 'The autobiographical process', *Current Sociology*, 43 (2-3) (1995), 161-177.
Campelli, E., 'Le storie di vita nella sociologia italiana: un bilancio', *Sociologia e ricerca sociale*, 31 (1990), 179-195.
Combe, S., 'Mèmoire collective et histoire officielle', *Esprit*, X (131) (1987), 36-49.
De Gaulejac, V., 'La socioclinique: roman familial et trajectoire sociale', in G. Pineau and G. Jobert (eds), *Histoire de vie. Approche multidisciplinaire* (Paris: L'Harmattan, 1989), pp. 25-38.
Ferrarotti, F., 'Sull'uso delle storie di vita nella ricerca sociale', in M. I. Macioti, *Biografia, storia e società. L'uso delle storie di vita nelle scienze sociali* (Napoli: Liguori, 1997), pp. 21-41.
Godard, F., 'Biographie et cycle de vie', *Enquete, Cahiers du CERCOM*, 5 (1989).

Kaha, C., 'Memory as conversation', *Communication*, XI (2) (1989), 115-122.
Lewis, O., *I figli di Sanchez* (Milano: Mondadori, 1966).
Passeron, J. C., 'Biographies, flux, itinéraire, trajectoires', *Revue Francaise de Sociologie*, XXXI (1990), 3-22.
Ricoeur, P., *Tempo e racconto* (Milano: Jaka Book, 1983, repr. 1986).
Shaw, C. R., *The Jake-roller, a delinquent boy's own story* (Chicago: Cambridge University Press, 1930).
Thomas, W. I., and F. Znaniecki, *Il contadino polacco in Europa e in America* (Milano: Ed. di Comunità, 1920, repr. 1968).

Liana M. Daher

2.
The Consequences of Social Movements. Methodological Links Between Sociology and History

How can we develop operative approaches making use of the multifaceted relationship between history and sociology methods? How can we make use of both these methodological perspectives in understanding social change? Can the sociologist combine "pure" historical ideographic and sociological qualitative methods in researching social conflicts and social movements in the contemporary world? Are these methods suitable to successfully investigate the long-term results and consequences of social movements in complexity dimensions?

The study of history assumes a key role in research on social movements for several reasons. The main reason focuses on the processes – occurring over time – activated by social movements that produce effects and/or changes both on the lives and stories of the individuals and groups participating in the movements, and on the flow of everyday life, opening up some possibilities and trends and closing others, promoted or hindered by elements deriving from different contexts (Markoff 2015: 82). Therefore, the historical method can be considered useful in capturing empirical knowledge on the effects of social movements, analyzing and evaluating their impact on contemporary society.

Indeed, understanding social change is a very complex matter; moreover, researching and isolating the cause and effect link in each single social change is one of the biggest challenges of sociology. If the researcher's goal is to highlight the link between social change and collective movements, then not only the events/processes deriving from historical and social decisions, actions, and strategies of collective movements should be considered, but also the interrelationships between movements and socio-historical processes that may give rise to independent and "dialectic" changes.

As we will see further on, the collective action of social movements could generate and/or combine with other social subjects to effect key changes in social structures and human life. These results could be considerably different from the intended and foreseen purposes of leaders and activists of social movements. Moreover, results often conflict with the stated intentions of the activists of the movement, producing unanticipated consequences – mostly in the long range – in the social structural conditions and life dimensions to which they were addressed.

Knowledge of the processes through which events and results have been developed – that would enable understanding of the original goals of the collective actions and strategies of social movements – would be useful for the analysis of results and consequences and to outline connections among the three principal time registers of the analysis: past, present and future. This would provide in-depth sociological understanding of the processes of social movements, and their effects on the world.

As noted by Weber (1922), the problem of historical causation is to determine the role of antecedents in the origin of the events. This socio-historical reconstruction implies several phases[1], but it is not exhaustive and is based on the relationship among *some* aspects of historical individuality and *some* prior events (Aron 1989: 467, emphasis added by the author).

The human socio-historical reconstruction can, of course – through the memory of prior events – provide key details. The case we are going to discuss is a reconstruction aimed at the specific cognitive interests of exploring complex collective dynamics, such as those of social movements, that produced substantial short and long-term changes on social life, even if in interaction with other processes, activated by other social subjects. Following Weber, the above socio-historical causation processes can be expressed in terms of probability or chance.

2.1 *Historical methods and sociological analysis: oral history in social movements research*

Original historical methodologies, such as oral history and archive research, seem to be increasingly used as tools to gain knowledge of the

---

[1] The four phases described by Weber are: a) construction of the historical individuality, exactly highlighting the characteristics of the event whose causes are sought; b) complex analysis of the historical event, stressing all its distinctive elements; c) if we are analyzing a unique sequence of events, it will be better to assume a different sequence of the events, answering the question: what would have happened if...; and d) compare the real sequence of the events to the imaginary development to identify the concrete elements that caused the socio-historical event subject of the analysis (see Aron 1989: 466-7).

social movement as a process-event[2]. There is a substantial difference in the way that which historians and sociologists achieve their scientific results: while historians see their findings as perfectly situated in time and space, sociologists try to go beyond space-time coordinates in their analyses (Goldthorpe 1991). The difference is not just related to the nature of the source material from the past, but in the different style of analyzing and understanding data, even if they are collected by the same method. Moreover, the frame in which these data are situated presents fundamental differences in terms of theory and paradigmatic references.

> Oral historians document the past by preserving insights not found in printed sources. […] oral histories can provide insights not normally found in more traditional reviews or summaries. […] The interview process practiced by oral historians affords participants in historical events an opportunity to address the historical record directly, to clarify what they see as misconceptions in third-person accounts, to discuss their own motives and those of other participants, and to provide their own personal assessment of the significance of the events in which they took part. This approach makes possible a clearer understanding of the intent of the participants than could be inferred from a record of the events alone (Russell n.d.: 1).

We can find remarkable affinities between the oral history methods, typical of historical sciences, and the sociological narrative interviews, particularly concerning the methodological correctness and aims in the first phases of the investigation in the historical approach, as noted in the quote above. The "oral history" method – both in the extensive modern use of the term and in the old meaning[3] – can be sociologically considered as a *collaborative process of narrative construction* (Nagy Hesse-Biber and Leavy 2006: 152). The strength of this process is «to allow individuals to share their story, hear their voices, and minimize the power relationships that often exist between a researcher and interviewee» (Creswell 2007: 40), with the final goal of acquiring complex knowledge directly from people with certain attributes of life experiences.

---

[2] In particular, oral history has been a technique usually applied by social movement scholars to analyze the subjectivities, individual and collective perspectives pertaining to the social movement organizations with the aim to capture memory of the past and reconstruct social movement socio-historical processes (among others see Fraser 1988; Passerini 1988, 2004; Portelli 1990; Grele 1991; Abrams 2009; Bonomo 2013).

[3] «Oral history is as old as history himself. It was the first kind of history» (Thompson 1978: 25).

Recent studies on social change have shown this approach to be suitable and useful in the reconstruction of time processes (McLeod and Thomson 2009; Daher 2013): it is possible to reconstruct sequences, within which the behavior of social actors acquires meaning, through the subjects' narratives that contribute to the description of the developing processes and condition of past events. In this sense, the longitudinal vision of the process *goals-collective actions-results* of social movements can be outlined in two different time perspectives: *through memory* – collecting the narratives of what happened in the past, while acknowledging that the truthfulness of the interviewees' statements depends on what they remember and/or want to remember or tell (Bichi 2002: 53) – , and/or *analyzing legacies*, i.e., asking for a comparison between the objectives of the movement, as personally remembered and perceived, and long-term results – a temporal perspective that shifts the narrative to the present. Consequently, stories and memories are included in a temporal process and, if the research aims at analyzing narratives of a quite conspicuous purposed sample of activists, this analysis will contribute to building a longitudinal dimension of the objectives-outcomes-consequences of the investigated movement in order to stress the past-present and future connections and understandings, and to point out the impact of social movement collective action on present (and future) times.

The biographical-narrative perspective will use, in this sense, the experience *told*, where remembering, forgetting, and building are the three main dimensions of the narrative (McLeod and Thomson 2009: 41). Through these dimensions, the subjects will be able to reconstruct sequences of events within which their behavior, and that of other members of the movement, acquire meaning, and contribute to the description of the previous socio-historical events and conditions through the above reconstruction process at a personal and collective level.

Of course, the oral history method applied to social movement sociological research should consider several structural questions, particularly concerning the intentional and unanticipated results and consequences of social movements. This is why some key issues should be outlined before examining the methodological subject.

As noted by Arendt (1958), every action has *multiple* consequences that could be set in an *unlimited* range of time. Extending this assumption from the single to the collective actions, we can understand how the socio-historical processes of the outcomes and consequences of social movements are difficult to rebuild, also because their actions are situated in a context of social plurality. In such a context it is not always possible to attribute outcomes exclusively to the action of social movements, i.e., it is difficult to draw a single direct casual connection between the specific goals of the movement and institutional and cultural changes.

Moreover, as noted by Sztompka (1991), direct results could change or be lost during the *long-term developments*. Applying this assumption to the social movement analysis, in the long term their collective actions could generate and/or combine to effect crucial modifications in society and human relationships, that are considerably different from the intended and foreseen purposes of leaders and activists of social movements. Often results conflict with the stated intentions of the activists of the movement, producing unanticipated consequences both in the life dimensions they were addressed to, or in other areas[4].

2.2 The engagement of sociology and history in the collection of memories of movements

As outlined above, when the main research goal is to rebuild a longitudinal dimension of the link between collective action and outcomes, two different temporal perspectives can be used: *memory*, regarding what happened in the past, and *legacies*, i.e., the exploration of the long-term outcome of the previous event. Historical methods can help us to operatively reach the above goal. Before putting forward a proposal of integration between sociological and historical methods – stressing the effectiveness of oral history in achieving results in the investigation of social movement consequences following the two perspectives just outlined – some points have to be addressed following the previous argumentation.

As stated by Mills (1974), without referring to the history, social science scholars cannot put forward social problems. In order to clearly express *what-needs-to-be-explained*, we should know as much as possible about the potentially complete information that must be provided only by the knowledge of the historical strains of human society. In order to avoid a mere and flat description of the social events, the variety of historical and contemporary social structures should be studied: historical types are

---

[4] This assumption recalls Merton's concept of unanticipated consequence as outcomes of a purposeful action that are not intended or foreseen. As Merton outlined, the concrete consequence of a purposive action may produce change: directly to the actor or to other people through: (1) the social structure; (2) the culture; and (3) the civilization (Merton 1936: 895). This argument can be easily applied to the understanding of social movement consequences.

fundamental in our investigation and comprehension of the social world. Each society frame is specifically historical; the knowledge of this frame will allow the social scientist to move toward a dynamic investigation of social change.

This is why sociologists should take history seriously. We can count at least three ways in which history matters. As stressed above, firstly, events happen within a historical context, which influences the events processes; secondly, behaviour, attitudes, and strategic choices take place in specific cultural contexts situated in a temporal-space range, so in a definite historical time – it is within this context that actors can learn from experience; and thirdly, history is not a chain of independent events, i.e., events that happen in time A could have important consequences for time B (Steinmo 2008: 127-8). The three reasons why history matters in the sociological understanding of social events and problems strengthen the issues we have discussed above regarding the investigation of the consequences of social movements. In order to study and rebuild events and/or social processes – as the social movement can be defined[5] – it is necessary to explore the historical frame and to consider that actors (interviewees) offer narrations enriched with historical elements. Therefore, carefully outlining the historical times in which the events took place will be a good frame and a useful tool for the sociological understanding of the research object.

The second point concerns the process of collecting individual-collective memory: studying the memories of specific events involves the collection of several narratives comprising what we define collective memory.

Oral history recognizes that memories are contingent and fluid; it is not just a method to gather information about previous events (Bosi and Reiter 2014: 131), but a creative means to reconstruct personal experiences of the past through the present interpretation, also in terms of meanings, given by the subject involved[6]. This is certainly positive for the richness of the narrative, but not always useful to obtain an account corresponding to what really happened then. As stressed above, the truthfulness of the interviewees' statements depends on what they remember and/or want to remember or tell. Indeed, the interpretation/remembering process and the correspondence with real fact is an issue that has to be considered by social scientists using the oral history method. The quality of research

---

5   The movement can be broken down into possible collective phenomena and forms of collective action through which it occurs, differentiating the aspects that do not belong to the social movement, that is rebuilt as a unity by the researcher following the purpose of the investigation (see Daher 2012: 30-1).
6   See Passerini 1988; Grele 1991; Portelli 1990; Abrams 2009.

data, the use of a hermeneutical perspective in conducting interviews (and in the analysis of narratives), the continuous movement of dialogue, and the continual and reciprocal interlocking of questions and answers make it problematic to define both the *congruity* of the different interpretations and the *conformity* between experience and narration. Consequently, we must take into account the limits and strengths of the method, giving an interpretation based on a hermeneutical sensitivity of the researcher (interviewer) who "knows that he does not know", but helps the other (interviewee) to reach knowledge of himself (Montesperelli 1998: 29).

As noted in a previous analysis on the topic (Daher 2015), memory is not exactly a mirror of the past, but rather a set of traces, footprints, and clues, which ask to be interpreted (Eco 1973), where individual recollections comprise the main element through which memory unfolds and rebuilds itself. Often, it becomes shared culture denoted with the expression, both widespread and challenging[7], of "collective memory", designating the social framework that guides and strengthens individual recollections through legitimization strategies.

Sociological research establishes the link between societies and their past; its main interest is not to capture events meticulously, but rather to use memory and recollections for the interpretation of the present and the understanding of social phenomena in the contemporary world. The representation of the past and future intertwine: on the one hand, memories are influenced by the interests and current projects of individuals and groups; on the other, memories influence the way in which the future can be anticipated and concretely transformed into action. This can be seen at all levels of social life: cultural, economic, legal, political, and technological.

A link emerges between the way we represent our past and our horizons of expectation; this interdependence, also oriented by the personal and social interpretation of the concept of time, opens up new perspectives in the field of sociology. It is evident that memory suffers from a whole series of disadvantages deriving from numerous and possible subjective interpretations of the same event, as well as from the subjective reworking of memories. Therefore, the idea that past experiences and events can be individually and collectively recorded and then faithfully recalled must be rejected in favour of a conception of memory as a social construct and continuously updated elaboration of the past (Halbwachs 1950).

---

7   In this regard, we are referring to the possible hypostatization of the concept which must instead be defined through the sharing and human interaction operations starting from the individual subjects (see Daher 2002, 2020).

This latter interpretation refers to an *intersubjective* dimension of memory: shared among multiple subjects, and communicated among subjects. It is not an "ability" held by an individual independently of the others, but an exchange-subject to which a lot of people refer (Jedlowski 1991: 27).

The comparison between past and present and an approach based on the detection and interpretation of memory can certainly support the researcher in the reconstruction of the aforesaid temporal nexus, also in relation to the aforementioned differentiation needs. In this case, memory will be considered as a *resource* contributing to giving meaning to reality, and to the relationship between continuity and change (Montesperelli 2003: 106-112).

This role of memory emphasizes the connection between sociology and history, at least the sociology which diverges from pure standardized approaches, just aimed at identifying recurrences, turning instead to non-standard – and tendentially idiographic – methods focused on understanding phenomena in their uniqueness, putting aside statistical representativeness, even if this sociology remains within social complexity conditions.

However, memory does not perfectly coincide with history. History stays still, becomes fixed in books and documents, and it is placed above and outside human groups; unlike history, memory is unstable and in constant change (Halbwachs 1950: 118-121). Rather, memory contributes to humans' participation in historical transmission by being "the experience of the past": the present employs the past through memory, to critically reappropriate the tradition (Montesperelli 2003: 118).

2.3 *On the effectiveness of oral history in sociological combined methods: a proposal*

In the light of the above considerations, it seems beyond doubt that oral history can be useful to collect memories of what happened in the past even in a sociological understanding. It will be useful, at this stage, to point out why a method that asks the witnesses to talk about the past, present, and future in a narrative way fit our needs in investigating the long-term results and unanticipated consequences of social movements.

The idea is to rebuild the longitudinal dimension of the link collective action-outcomes in two different temporal perspectives, giving information of what happened in the past (memory), and exploring, through the narrative of the same subject, the long term outcomes/consequences of social movements (legacies). The proposal is a combined method, already empirically tested[8], aiming at investigating personal and identity paths related to the participation in the movement, and reconstructing reasons and goals of the collective action in the past. This is in order to compare the results of investigation to the present or recent past and to understand if the movement has achieved its stated goals in the long term or if backlashes and/or unanticipated consequences developed.

Making use of a narrative method (McLeod and Thomson 2009) – as useful for rebuilding temporal processes that does not claim to represent a close and detailed model of the above change processes – seems to be particularly suitable because of its implicit power to give an analytical frame of the past, the present, and the future as constitutive and inseparable parts of the temporal flow (Elias 1992). These research methods, data, and relative interpretations are characterized by a dynamic relationship between temporal registers where it is possible to distinguish the above dimensions. This view plays a fundamental role in the choice of the approach and the method, as well as in the choice of guiding concepts for the interviews.

Taking advantage of what has already been highlighted, both in terms of the relationship between sociology and history and concerning the usefulness of oral history in the study of the consequences, primarily in the long term, of social movements, the first part of the interview was based on the *oral history method* (Nagy Hesse-Biber and Leavy 2006), to collect information/memories about the past in terms of the historical goals of the movements and the debate around the theme. The second part had the structure of a *focused interview*, to gather the interviewee's opinion on the congruence between the intended goals/purposive collective actions and the outcomes/consequences concretely achieved; in this last phase of the interview the principal aim of the researcher will be to encourage a meticulous narration of the interviewee's way of thinking (Bichi 2002), in order to directly gain complex knowledge from the participants/activists of the movement itself.

As pointed out above, oral history acts a *collaborative process of narrative building* (Nagy Hesse-Biber and Leavy 2006: 152): the individual takes part in the recounting of events as part of his/her personal history, but at the same time he/she is part of the historical and social events. During the oral history session, the subject is asked to speak about the way in which he/she was involved in the movement and his/her personal experience (*delivery*), from which we may deduce the shared ideas and objectives of the group,

---

8   See Daher 2013, 2018.

i.e., the "reasons-goals" of the protests. In this first section of the interview, after a first step of "free" narration, also helpful to involve the activist in the subject area of the conversation (Olagnero and Saraceno 1993: 91), sub-topics were discussed concerning the different outcomes deriving from the movement, targets partially or fully achieved, backlashes, and unintended consequences. Needless to say, the oral history interview step can be considered as a narrative, i.e., «an account of a sequence of events in the order in which they occurred to make a point» (Polletta and Gharrity Gardner 2015: 535), but much more than this is combined in this proposal. In addition to what has already been noted, the aim is first to collect, through the interviewees' accounts, the "share capital" of the movement that is the subject of the research. Moreover, oral history allows people to start with a "free" account of personal experience that was very helpful to start the conversation and to focalize together (interviewer and interviewee) the issues of the talk; this phase was also considered by the activists involved as a "cathartic way" of narration.

Passing from *memory* to *legacies,* not only do the contents change, but also the style of the interview. A focused interview could be a good method to gain information from an individual on a particular topic because it is *issue oriented.* It is planned to capture – through the narration – the interrelating dimension of past-present-future and their indivisibility, and specifically the dynamic relationship between the past and the present, characterized by both determination (*the past shaping the present*) and hermeneutics (*the present constructing the past*) (Connerton 1989: 61). Remembering, forgetting, and constructing were the three dimensions that occurred in the interviews; every witness was entreated to build a sense of the larger context and structures in which recollected experiences occurred, leaving aside or trying to recognize individual psychological processes and emotions.

Speaking about long-term and unanticipated outcomes, the researcher and interviewee returned to the present passing through the past. The memories helped the witnesses to compare the concrete targets – that were shared with the other participants of the movement – to the outcomes actually achieved, not only those from an economic or juridical point of view, but particularly those stemming from cultural perspectives, asking the question: what aspects of the widespread culture and behaviour of today coincide with yesterday's goals? This is the principal reason why the interview is structured to take advantage of a comparison between past and present format, and the guiding concepts are mainly related to the reconstruction of the connection between movement-collective action and its short and long-term outcomes.

Concluding remarks

The combined method of oral history and focused interviews bring an explicit biographical focus to the research on social change; in particular, it offers the way to know how social change events and occurrences are experienced at the individual level, and to implement past experiences as an interpretation of the present. Narratives and memories of interviewees could be read like a *journey*: «Constructed in the present, evoking and even transforming the past, and often told with a view toward the future, toward generational inheritance and a sense of other possibility» (McLeod and Thomson 2009: 53). The inter-subjective dimensions of the narration have to be read not only as a glance at the past; the memory of what happened then and the personal interpretation of it has, in our case, the function of a connection between the past and the present; moreover, remembering is also a way of looking at the future, trying to produce understanding starting from the previous experiences. The tricks of remembering and forgetting reveal the interplay between psychological and cultural processes in the social production of memory.

The combined method proposes to mix two different temporal lenses of observation focusing, first, on the rebuilding of the historical context, events, and processes, with the aim to focus, in the second phase, on the problem under investigation in the present times. This combined model of narrative interview aims at taking advantage of "naturally occurring" data to rebuild the sequences (*how*) where the meanings of social actors (*what*) developed, and consequently to specify the attributes of the events. In this model the yesterday's objectives of the movements are used as a *lens* for understanding the present and the direct and indirect, short-term and long-term results of the movement by the dynamic relationship between the past and the present outlined above.

As noted by Goldman, sociology and history have studied the same phenomena since as far back as 1974, each one returning only a theoretical and incomplete picture until the one integrates the contributions of the other, offering a *synthesis* of the two sides. Indeed, both the remarks and the proposals in this contribution go toward this possibility.

Bibliographical references

Abrams, L., *Oral History Theory* (New York: Routledge, 2009).
Arendt, H., *The human condition* (Chicago: The University of Chicago Press, 1958).

Aron, R., *Le tappe del pensiero sociologico* (Mondadori: Milano, 1989).
Bichi, R., *L'intervista biografica. Una proposta metodologica* (Milano: Vita e Pensiero, 2002).
Bonomo, B., 'Presa della parola: A Review and Discussion of Oral History and the Italian 1968', *Memory studies*, 6 (1) (2013), 7-22.
Bosi, R., and H. Reiter, 'Historical Methodologies: Archival Research and Oral History in Social Movements Research', in D. della Porta (ed.), *Methodological Practices in Social Movements Research* (Oxford: University Press Oxford, 2014).
Connerton, P., *How Societies Remember* (Cambridge: Cambridge University Press, 1989).
Creswell, J. W., *Qualitative Inquiry and Research Design: Choosing Among Five Approach* (Thousand Oaks: Sage, 2007²).
Daher, L. M., *Azione collettiva. Teorie e problemi* (FrancoAngeli: Milano, 2002).
— *Fare ricerca sui movimenti sociali in Italia. Passato, presente e futuro* (FrancoAngeli: Milano, 2012).
— 'From Memory to Legacies. Cultural Outcomes, Success and Failures of the Feminism Movement in Sicily', *International Review of Sociology*, 23 (2013), 438-460.
— 'Sociologia della memoria: strumento per l'interpretazione del passato, filtro per la comprensione del presente', *SYNAXIS*, XXXIII (2015), 79-95.
— 'Le *unanticipated consequences* dei movimenti sociali: Considerazioni teoriche e relative alla ricerca sul campo', *Studi di Sociologia*, X (2018), 1-18.
— *Azione collettiva. Teoria e ricerca empirica* (FrancoAngeli: Milano, 2020).
Eco, U., *Segno* (Milano: Isedi, 1973).
Elias, N., 'Time: an essay', in N. Elias, S. Mennell and J. Goudsblom (eds), *On Civilization, Power, and Knowledge. Selected Writings* (Chicago: University of Chicago Press, 1992), pp. 253-268.
Fraser, R., *1968: A student Generation in Revolt* (London: Pantheon, 1988).
Grele, R., *Envelopes of Sound: The art of Oral History* (New York: Praeger, 1991).
Goldman, L., 'Il pensiero storico e il suo oggetto', in P. Crespi (ed.), *Sociologia e Storia* (Milano: Celuc, 1974).

Goldthorpe, J. H., 'The Uses of History in Sociology: Reflections on Some Tendencies', *The British Journal of Sociology*, 42 (2) (June 1991), 211-230.

Halbwachs, M., *La mémoire collective* (Paris: Presses universitaires de France, 1950).

Jedlowski, P., 'Il testimone e l'eroe. La socialità della memoria', in P. Jedlowski and M. Rampazi (eds), *Il senso del passato. Per una sociologia della memoria* (FrancoAngeli: Milano, 1991).

Markoff, J., 'Historical Analysis and Social Movements Research', in D. della Porta and M. Diani (eds), *The Oxford Handbook of Social Movement in Western Europe. A Comparative Analysis* (Oxford: Oxford University Press, 2015), pp. 68-85.

McLeod, J., and R. Thomson, *Researching Social Change* (London: Sage, 2009).

Merton, R. K., 'The Unanticipated Consequences of Purposive Social Action', *American Sociological Review*, 1 (1936), 894-904.

Mills, C. W., 'L'uso della storia', in P. Crespi (ed.), *Sociologia e Storia* (Milano: Celuc, 1974), pp. 191-216.

Montesperelli, P., *L'intervista ermeneutica* (FrancoAngeli: Milano, 1998).

— *Sociologia della memoria* (Roma-Bari: LaTerza, 2003).

Nagy Hesse-Biber, S., and P. Leavy, *The practice of qualitative research* (Thousand Oaks: Sage, 2006).

Olagnero, M., and C. Saraceno, *Che vita è. L'uso dei materiali biografici nell'analisi sociologica* (Roma: Carocci, 1993).

Passerini, L., *Storia e soggettività: Le fonti orali, la memoria* (Firenze: La Nuova Italia, 1988).

Passerini, L., *Autobiography of a Generation Italy, 1968* (Middletown, CT: Wesleyan University Press, 1st ed. 1996, repr. 2004).

Polletta, F., and B. Gharrity Gardner, 'Narrative and Social Movements', in D. della Porta and M. Diani (eds), *The Oxford Handbook of Social Movement in Western Europe. A Comparative Analysis* (Oxford: Oxford University Press, 2015), pp. 534-548.

Portelli, A., 'Intervistare il movimento: Il '68 e la storia orale', in P. Poggio (ed.), *Il Sessantotto: L'evento e la storia*. (Brescia: Fondazione Luigi Micheletti, 1990), pp. 125-132.

Russell, D. E., (n. d.), *Oral History Methodology, The Art of Interviewing.* Available on: http://marcuse.faculty.history.ucsb.edu/projects/oralhistory/199xDRussellUCSBOralHistoryWorkshop.pdf

Steinmo, S., 'Historical institutionalism', in D. della Porta and M. Keating (eds), *Approaches and Methodologies in the Social Sciences*, (Cambridge: Cambridge University Press, 2008), pp. 118–138.

Sztompka, P., *Society in action: The theory of social becoming* (Chicago: University of Chicago Press, 1991).

Thompson, P., *Voice of the Past: Oral History* (Oxford: Oxford University Press, 1978).

Weber, M., *Gesammelte Aufsätze zur Wissenschaftslehre* (Tübingen: Mohr, 1922).

PAOLO DE NARDIS

3.
THE CONSTRUCTION OF SOCIAL ACTION.
FROM THE POSITIVISTIC TO THE VOLUNTARIST THEORY OF THE ACTION

Antonio Labriola's theory on the social action of his times tries to overcome the positivistic approach, thanks to a critical review of the political sciences. As we know, Labriola's support of a materialistic interpretation of history and the social actor is based more on an empirical approach than an idealistic one.

His description of social phenomena focuses on the morphological and nomological features. It does not pertain to early positivism and its mechanistic metaphysic, but it seems to underline the need for empirical norms for the explanation of the social facts. This is clear in his description of the social group dynamics (supported by the empirical social psychology of Johann Friedrich Herbart) and social collectivity, interpreted as the social actor.

However, Labriola's overcoming of positivism does not correspond to the systematic development of sociology as a specific discipline, which happened with Durkheim and Weber in France and Germany.

It is evident that from 1870 to 1885 Labriola sympathized with the German cultural trends involved in the rising debate on the relationship between *the sciences of nature* and *the sciences of the spirit*. This experience could explain why Labriola embraced Herbart's approach and empirical psychology to build a new method for the *normological explanation* of historical facts. Labriola's introductory lecture at Rome university in 1887 *I problemi della filosofia della storia* (the problems of the philosophy of history) is a detailed work on the real evolution of the historical processes, their formalization and their *rules*; so, in brief, it is an analysis on the possibility to consider history as a science, albeit constitutionally different from the sciences of nature.

First of all, it is important to explain how Labriola describes the differences between *principia cognoscendi* and *principia essendi* in this text: the former concern what is connected to the method of historical research, and the latter concern concrete reality and the consistency of historical facts. In this way, the author intends to clarify the problems regarding "the general systematic" in which we can place and manage the historical process.

The reflection on the "objectivity of the exposition" is explained through some theoretical coordinates thanks to some contents already used by historians, philologists, linguists, and also some of Herbart's followers like Steinthal, Lazarus, Droysen, Boeck. The first result of this analysis is that the certainty in the research of the historian "is not measurable only with the instrumental precision of paleographic, philological, and linguistic methods, but also – and primarily – thanks to the level of theoretical transparency and the reproducibility of the analyzed subject"; therefore, "the theoretical elements used for the interpretations of historical facts, when they are adequately explained, lead to general disciplines, that are the cornerstone of each further analysis"[1]. In other words, in order to understand the events, you cannot refer to a general form of inductivism (facts do not explain themselves), but it is important to have some theoretical frameworks.

The two main analytical themes for Labriola are the *epigenetic theory* on one hand, and the *morphologic theory* on the other. Indeed, his first presentation of the historical processes highlights "specific forms of relationships and overview" which explain the "nexus and plexus" of the human activities that are "resistant forms" to the corrosive action of the contingencies that aim to persist".[2]

Labriola believed that there is not a single principle that enacts these activities in a determinist way, and in his work he strongly criticizes the monist and evolutionist vision of history; therefore, even if the birth and the development of new disciplines such as anthropology, ethnology, and sociology substantially extended the factors that build history, many of the visions of the historical process share a form of monism and holism and risk falling into the trap of Hegel's ontology or Schelling's equal absolute.

Therefore, Labriola persists in focusing on the *destruction*: this can clarify the fact that historical processes show the development of human activities toward a clear *epigenetic theory* of the social-historical structure.

The main influence on these theories came from the German debate on the basic cornerstones of the *sciences of the spirit*. From this point of view, the influence of scholars focused on social analysis, like Schäffle, favored the end of the proto-hegelism of Labriola.

---

1 Labriola, A., *Scritti filosofici e politici*, ed. by F. Sbarberi, Torino, Einaudi, 1973, p. 9.
2 *Ibidem*, p. 15.

In addition, the interest in the new perspectives joined the key roles of social action and human activities to the foundation of a science of society; therefore, the end of the dominance of philosophy in this respect gave rise to the relationship between history and sociology. As we can see, this was the ultimate defeat of positivism by German philosophy at the end of the century, with its critics of evolutionist monist determinism, albeit through concepts borrowed from empirical psychology and, particularly, by social psychology.

In a review, Labriola and a follower of Herbart discuss the methodological fertility of social psychology, that starts from the psychology of the individual and moves across the *genetic method*, considering the basis of psychological research in the debate on the principles of the new science of psychology. In this debate, the psychic dynamic is characterized by the connection of many conditions and interests and not by a simple mechanic cause.

A specific value is derived from the concept of "law" for the historian: he or she cannot obtain this concept from the order of the extrinsic chronology of the facts because it is impossible to predict facts in this way. The concept of "law" has to be defined with the genetic-morphological method. It is a comparative method for the study of languages, myths, habits, and similar matters. It is the morphology of the organic sciences that suggested this comparative method and recognized the mutual condition of the action that produces a specific model; therefore, it is this morphology that transcends empiricism or general inductivism based on the "aggregation of unlimited news", clearly explaining that homologies allow a tradition or an institute to be completely defined, even if it is very old, thanks to the comparative method.

In this way, "the designation of the *genetic conception* seems to be clearer" and "more comprehensive" because it can comprehend not only the real content of things, but also the logical-formal ability to understand how they come into being. So, the expression of the "genetic method" does not touch the empirical nature of each formation, and, at the same time, defends the right control of the formal structures which make up the theoretical bases for the explanation of the facts.

All these concepts give strength on one hand to gnoseological objectivism, and on the other hand to each aspect of gnoseological teleology which is against the preconception that things are related to a rule, an idea or a goal – explicitly or implicitly. The problem is the reduction to "a series of explanations" of the conditions and the variables of historical processes, giving parameters for an analysis and evaluation of historical change.

Against linear determinism there is the necessity, but at the same time, the obstacle of the form of prediction by Engels and, to some extent also by Marx, in which the "forewarnings and promises" are replaced by the "morphological" prevision, because such a prediction is possible and not apodictic or deterministic.

Lazarus and Steinthal were crucial reference points for Labriola's revision of the distinction between the "sciences of nature" and the "sciences of the spirit". Boeck, Doysen and, above all, Paul were other reference points for Labriola for some other aspects connected to the scientificity of historical research. In particular, they inspired Labriola regarding his criticism of the all-embracing vision of history, starting from the development of some reflections on the psychology of communities by Lazarus and Steinthal.

The real originality consists in the "epigenetic theory of civilizations" based on the interaction between *physical* and *psychical forces*: from this interaction, "forms" of civilization could emerge like the one described by the psycho-physiology of Wundt. He argued that all sciences – including psychology – had to utilize the historical methodology: thus, the Wundt psychological method became a fundamental tool for historical analysis.

From these assumptions Labriola inferred that the psychological processes and their searching for rules could be very beneficial for the clarification of the social phenomenology perspective. Moreover, the questions regarding the individual and social action structures could be derived from the same reflections. Labriola was very interested in exploring the mechanical or organic structure of society, its formations, and social development.

In this perspective, a certain hostility between the historical sciences and the social empirical ones is still evident today, with a specific reference to sociology. This makes it difficult for scholars of the two different disciplines to collaborate and gives support to the idealistic premise that the historical event is unique.

It seems a very old issue, typical of the early *Methodenstreit* when Windelband tried to theorize the already ancient distinction between idiographic and nomothetic sciences. However, it often reappears and what is worse is that it is important for sociologists who are ready to point out the evident difference between history and sociology. Therefore, it is tantamount to saying that Weber's research on the relationship between Protestantism and capitalism, carried out with a rigid nomothetic method as for *historical sociological research*, is not possible (or not yet).

Is it the time to return to the reflections on history and sociology and the opportunity to identify a unique method shared by all human sciences – a nomothetic method for the historical-social sciences? Perhaps the time has come because it is a very stimulating question and very relevant today.

There is an all too frequent assumption that sociologists – especially those that tend to be more theoretical – avoid history or the philosophy of history or the theory of history. This is true if we consider the fact that they refuse any absolute concept in terms of any opportunistic use in the research. Therefore, undoubtedly, when they work on a question such as *social change,* it is not possible to exclude the *theory of the history*. However, if social change requires a scientific theory, it can be done only through the various *sociological* theories on this process; also the theory of history in a scientific context is a theory of the theories of history: therefore, it is the theory of those whose works concern this or that aspect of history: it is a *theory of historiography.* We can infer that, in other words, a theory of social change can, generally speaking, correspond to a theory of historicity, even if it cannot be a more or less direct consequence of a theory of historiography. From this point of view, we can note that sociologists can learn from the metahistorical, even if, although we should not be surprised that sociologists are not interested in it, it seems rather strange that historians often are, with the exception of scholars like Braudel and Le Goff, who are some of the "global" historians of the *Histoire Nouvelle* of the *Annales.*

However, if historians and sociologists deal with historicity, albeit starting from different points of view, it is not clear why there is this lack of mutual tolerance and complex dialogue. If we agree that the common question is the explanation of the historical-social processing on the one hand, and, on the other hand, the logical unit of scientific knowledge, perhaps we should consider that a theory of historiography that makes a theory of historicity possible may be feasible. Obviously, this would be carried out by a rigorous method of historical research, checking the assertions it is built on and therefore ensuring it is based on logic.

We could ask the question: why do historians not use a theory of social change? Should we believe that all historians share the idea that the scientific concept of theory is not adequate for the explanation of historicity? In brief, is it not possible to explain historical events as processes through nomological fundamentals and classifications regarding individual events?

The historian's explanation of an event refers to a general model, albeit not explicitly, even when it is not totally idiographic. Historians are required to put the explanation into a series of propositions that could provide a sense; otherwise, the explanation could appear weak.

Therefore, if history is change, the change is only measured in comparison with something constant that does not change. This unchanging condition asks for general propositions for the investigated project. Starting from this point of view, the theories of change concern the identification of the historical-social variables assumed, for hypothesis, which are important for the explanation of historical processing (a series of events). Only this premise can legitimize the research carried out by historical sociology.

As noted, the debate on the sciences of the spirit arrived in Italy thanks to some scholars. Labriola was one of the most attentive to this theme, even if the *Methodenstreit* with the publication of *Enleitung* by Dilthey was not understood, as it later emerged.

It is clear that in Labriola's proposal of a theory of the historical-social phenomena there is something that *overcomes* the simple historicism in the rational necessity (and not for the simple possibility) to resort to a *nomothetic* perspective, in other words through general rules in the production of historical knowledge and the human action.

It is important to note that in some of Labriola's methodological proposals, he is afraid of a simplistic use of the historical categories in the evaluation and goals of social research, so that he can avoid misunderstandings and misconceptions. Actually, the difficult application of the old historicistic categories to the "Sciences of the spirit" were becoming clear during the debate on German historicism. Some unsolved problems in the proposals of Dilthey, Windelband and Rickert were inherited and subsequently partially solved by Max Weber in the way he completely overcame the historicistic perspective. This excess of the old historicism can be seen in Weber's position regarding the problem of the relationships between history and sociology.

In Labriola's theoretical reflection, the interest in the problem regarding historical-social sciences, starting from the criticism of the voluntarist theory of the action on the one hand, and the path of the idealistic philosophy of history on the other hand, emphasizes the need for the explanation of the phenomena through a genetic-morphological perspective.

The criticism of irrationalism and empathic intuitionism meet the need for explanations and a meticulous searching for new theoretical categories that make the explanations easier. However, the debate on the sciences of the spirit arrived in Italy in some way and was followed with difficulty by some scholars who created an analytical methodology, even if they did not find ready interlocutors. Durkheim in France, on the one hand, and Weber in Germany, on the other hand, founded the discipline: Durkheim countervailed the "psychology" of Tarde, and Weber closed the first period of the *Methodensreit* with his eminent profile.

In the same period, in Italy, sociology did not exist, in spite of the wait for this new discipline and Labriola's attempts at the construction of an empirical social science that could explain society after the first attempts of a naive positivism.

In spite of these difficulties, Labriola focused his reflections on scientific explanation, because he believed that explicative reports of the historical-social phenomena (maybe also *predicitive* reports when not ideological as for the philosophies of history) are the fundamental result for the human and social scientist.

The logic of positivism (or of positivisms) and its naïve traditional dualism of the sources of knowledge (that is the intelligible opposition), largely produces a theory between naturalism and historicism and promotes a bipolarism without mediation between the instance of the explanations and the comprehension. They had been defined in 1868 by the historicist Droysen before a debate on the method that developed dichotomies like nature and nurture (spirit) for Dilthey, and nomothetic/idiographic for Windelband. They constantly evoked an old dichotomy from Romanticism like necessity/freedom that, later appeared in the dichotomy generality/individuality. This happened, clearly, because of the unsustainable position of early positivism in the XIX century and the weakening of the methodological debate on the sciences of the spirit that weakened Droysen's Manicheism.

In Labriola's thought, Droysen represented an essential tool for his anti-positivistic criticism of the foundation of a logic of scientific-social knowledge.

The profile of the social philosophies of Comte and Spencer was more similar to the philosophy of history than to science: they saw science as a gnoseological activity with a methodological foundation. This derives – and Labriola seemed to be informed of this, albeit implicitly – from the reaction of German philosophy during the second half of the XIX century: it was an immediate and legitimate reply to awkward attempts of perspectives and opinions scientifically oriented, albeit based on ideological positions. As is known, after Droysen and other scholars, in his *Einleitung in die Geisteswissenschaften,* Diltey destroyed all the approaches to the social sciences from the perspective of the philosophy of history.

From this point of view, German historicism should not be considered as a completed program for the methodology of historical-social sciences, but only for the study of historical events. It was a hard strike against the "scientistic" naiveties of the early positivism of the XIX century. This strike was considered, in an extremist way, as a strong opposition to positivism and some existentialist approaches that it inferred. These would be later placed in the phenomenological sociological orientation, and recently, in the ethno-methodological one.

A final point to be considered is the following: as we have noted, another German scholar important for Labriola's thought was Schäffle, especially for his analysis of the social organisms that evoked a holistic concept of the social. However, Labriola soon kept Schäffle's theory at a distance.

This "organic sociology" did not allow the separate analysis of a specific component of a wider system because its nature was characterized by its function in that context, as if it produced interactions with the other components of the system/organism.

Later, Weber would stress that the fundamental feature of this perspective had been drawn by Schäffle himself in his *Bau und Leben des Sozialen Korpers* (Tübingen, 1875-1878), that is one of the books that mostly oriented Durkheim's interest in sociology.

Schäffle's influence (like Wundt's) can be seen in the adoption of a genetic methodology for an explanation – not naively evolutionistic – of the historical-social process. According to his conceptual construction, the relationships among the individual, history, and society were identified in the evolution of the functions of the social organism with specific *psychogenetic* characteristics compared to a cultural change that connects individual and social levels.

An "epigenetic of civilization" theory can be inferred from all these reflections: a method that comes from the definitions of the historical formations and that moves to the *explanation* of their construction in the society inferred. It could enable everything connected to all the other things regarding conditions, premises, and inferences to be highlighted, but not only for a simple mechanic cause.

From this point of view, the most balanced methodological tool appears to be the empirical psychological one: it was the only discipline that caught the characteristics of the cultural phenomena as opposed to a purely mechanic reductionism. It represents the starting point of the theoretical bases of German historicism and the debate on the method towards a rigid scientific conceptualization of the social and a re-qualified sociological explanation. Everything was in an embryonic form, even if it was already clear, waiting for an empirical social science, epistemologically solid and methodologically able to make explicative-predictive reports on the social phenomena.

Bibliographical references

Berthelot, J. M., *La costruzione della sociologia* (Bologna: Il Mulino, 2008).

Bonolis, M., *Storicità e storia della sociologia* (Milano: FrancoAngeli, 2007).

De Nardis, P., *Aspettando la sociologia. Antonio Labriola dalla psicologia empirica alla spiegazione sociologica* (Acireale: Bonanno Editore, 1993).

De Nardis, P., 'Marxismo e protosociologia in Italia: un'analisi non positivista e antievoluzionista' in R. Vignera (ed.), *Neodarwinismo e scienze sociali* (Milano: FrancoAngeli, 2010), pp.145-170.

Labriola, A., *La concezione materialistica della storia* (Bari: Laterza, 1965).

Labriola, A., *Scritti filosofici e politici*, ed. by F. Sbarberi (Torino: Einaudi, 1973).

Lentini, O. (ed.), *La sociologia italiana nell'età del positivismo* (Bologna: Il Mulino, 1981).

Leonardi, F., 'Teoria e teorie del mutamento sociale', *Sociologia e ricerca sociale*, 19 (7) (1986), 5-34.

Leonardi, F., *Di che parla il sociologo? Problemi di epistemologia delle scienze sociali* (Milano: FrancoAngeli, 1986).

Poggi, S., *Antonio Labriola. Herbartismo e scienze dello spirito alle origini del marxismo italiano* (Milano: Longanesi, 1978).

Poggi, S., *Introduzione ad Antonio Labriola* (Bari: Laterza, 1982).

Gennaro Iorio

4.
Theoretical Thoughts and Historical Context: The Case Of Social Relationships

Introduction

All theoretical thoughts and scientific endeavors, even in their most abstract forms, are not only the product of intellectual reflection, but also the expression of a given society in a specific historical period. Since human beings live in society and all societies have a temporal dimension, cultural currents always have a precise historical and social context.

In this brief chapter, we have tried to highlight the fact that "social relationships" have been at the basis of sociological study ever since sociology was established as a science of human behavior. At the same time, the concept of social relationships created a boundary which distinguished it from philosophy, law, psychology, biology, economics, history, and politics, all of which concern the interpretation of social phenomena. Therefore, we propose to analyze the theories of those authors who, for the first time in the history of human thought, came to be defined as sociologists. The "discovery of society" presented by sociologists coincides with the individualization of new practices and new social relationships in the emerging modern society: therefore, the category of social relationships was invented at a theoretical level.

In making this cultural distinction, we used a method developed by that sector of sociology which looks at the development of knowledge. It considers the interdependence between theoretical models and the historical contexts in which they matured. Thus, the concept of interdependence prevents us from giving a single cause explanation to the relationship between social structure and social phenomena.

Having stated this sociological premise, I will begin to develop the theme on social interactions and their role in the origin of sociological reflection.

In proposing the discovery of social interactions as a focal point of sociological thought, we are certain to run the risk of structuring it excessively and oversimplifying it. This is due to lack of space to fully present this subject in all its facets.

4.1 *Acknowledging the timing and location of sociology*

Sociology is perhaps the only science which has an official starting date: in 1838, in the *47th Course of positive philosophy,* Auguste Comte (1798-1857) coined the term Sociology (Comte 1908, It. trans. 1967: 132). The writings of Comte had already constituted a point of reference for sociology in 1820. Since the first of these texts was written in collaboration with his teacher, Henry Saint-Simon (1760-1825), credit for the creation of this discipline must also be given to this writer.

Another pioneer was undoubtedly Herbert Spencer (1820-1903), a sociologist who lived in England during the Victorian era, in the mid-nineteenth century. Spencer had considerable influence on the history of social theory, and the first sociologists in the United States often referred to his writings – the United States was the nation where sociology gained its first strong foothold in academia.

Another important element in sociology is its historical development, i.e., the period in which it reached cultural maturity. This occurred during the years in which the founding fathers of this discipline were publishing their works – between 1880 and 1920: Emile Durkheim (1858-1917) in France; Georg Simmel (1858-1918), Ferdinand Tönnies (1855-1936) and Max Weber (1864-1929) in Germany; the Italian Vilfredo Pareto (1848-1923) who taught at Lausanne; and the American "founders", from Lester Ward (1841-1913) to Charles Cooley (1864-1929).[1]

Sociology began to be taught at university level in France and the United States; in Germany, it aroused interest in the academic world which led to the writing of the *Deutsche Gesellschaft für Soziologie.*[2]

The latter part of the nineteenth century saw the start of some important sociological journals: the *Revue Internationale de Sociologie* (1893), the *American Journal of Sociology* (1895), the *Rivista Italiana di Sociologia*

---

[1] The initiators of American sociology are considered to be: Lester Ward (1841-1913), William Graham Sumner (1840-1910), Franklyn Giddins, Albion Small (1854-1926), Edward Ross (1866-1922), William I. Thomas (1863-1947), Robert E. Park (1864-1944) and Ernest Burgess (1886-1966).

[2] Only with the beginning of the Weimar Republic would sociology be accepted by German universities. See F Ringer, *The Decline of the German Mandarins*, Harvard University Press, Cambridge Mass 1969, pp. 228 onwards.

(1897), and the *Année Sociologique* which Durkheim began to publish in 1898. In France the *Institute International de Sociologie* was founded, connected to their *Revue*, to which the most important sociologists of various nations belonged, with the exception of the followers of Durkheim.

Therefore, sociology arose as a discipline in the western hemisphere, particularly in Western Europe over the course of the nineteenth century and in the United States at the turn of the eighteenth and nineteenth centuries. I believe it is important to underline the fact that the origin of this science is linked to a specific context, namely its Euro-Atlantic roots. A second element to remember in this brief geo-historical map is the fact that sociology is historically positioned in a post-revolutionary era.

We have already mentioned that in France sociology was introduced by Comte and Saint-Simon in the years of the Bourbon restoration.

In England, Spencer carried out his studies not only after the revolution of 1688, but also after the parliamentary reform of 1832 and the abolition of the corn laws.

In Germany, Italy, and the United States sociology was formalized long after the decisive events of the bourgeois revolution.[3]

Therefore, it is unlikely that the emergence of a new science of society and social unrest can be attributed merely to chance. Moreover, sociology was not the first application of scientific methodology to societal life: political economics had already reached a state of maturity a century before the birth of sociology, and even earlier Hobbes (1588-1679) and Montesquieu (1689-1755) had already tried to analyze society with the methods of natural sciences (Nisbet 1977). The precursors of sociologists had to deal with two types of already-existing reflections on society: political economics and political theory, or rather, philosophy.

4.1.1 *The distinction from political economy*

Political economy did not arouse much interest in eighteenth century France; it occupied only marginal space in the writings of the first generation of sociologists. Comte and Saint-Simon were acquainted with and

---

3   In Eastern Europe there are Ludwig Gumplowicz (1838-1909 of Austrian-Polish descent), Tomàs Masaryk (1850-1937 from the Czech Republic), and from Russia Maksim Kovalevskij (1851-1916), J. Novikov (1901-1975), Evgenij De Roberty (1853-1915), Edward Westermack (1862-1939) and K. N. Michajlovskij (1852-1906). From a bit farther away there are the first Japanese scholars, among whom we recall Nagao Aruga (1860-1921) and Tongo Takebe (1871-1945).

appreciated the work of Adam Smith (1723-1790). They were influenced by the French economist Jean-Baptiste Say who endeavored to exalt the importance of industrial entrepreneurs rather than agrarian capitalists, who were in turn defended by the physiocrats. This is why Saint-Simon attempted to relate political forms to real social forces. His objective was therefore to place the power in the hands of the industrialists (Saint-Simon 1975). Comte, on the other hand, was opposed to an economic vision of society. He acknowledged the role of political economy in drawing attention to the new class of industrialists, yet he remained hostile to the narrow vision of social organization from the point of view of *laissez-faire* (Comte 1908, It. trans. 1967: 138-146). Even Spencer, while defending the laws of political economy from its adversaries, did not attribute any special importance to it in "A System of Synthetic Philosophy". He used concepts and arguments like the division of labor and trade developed by economists, but stated that his interest in this area came from physiology and that his point of reference was thus the science of biology (Spencer 1882: 334).

Hence the relationship between political economy and sociology does not seem to be a useful starting point to analyze the historical development of this new discipline. Its pioneers did not consider this new intellectual undertaking to be either a critique or a continuation of political economy.

4.1.2 *The distinction from philosophy*

The relationship between sociology and political philosophy is a different matter. The first sociologists spent a lot of energy considering the development of a political science and a political system that would correspond to the needs of the new world. As we have already seen, Saint-Simon based his entire reflection on the elaboration of a new political science capable of developing a political system consistent with the needs of the new world, which he interpreted as the building of an industrial order. Comte sought to express a positive science of politics, illustrated in a systematic manner in his *Plan des travaux scientifiques necessaries pour reorganizer la société*. In fact, one of his most important works is entitled *Système de politique positive*.

Alexis de Tocqueville (1805-1859), his contemporary and author of *De la démocratie en Amèrique,* also arrived at a conclusion from his own studies that «a new political science is necessary for a world that is now completely new» (Tocqueville 1968: 20). Spencer's intellectual efforts were not aimed at defining the development of a new political system inasmuch as his reflection was an integral part of his philosophy of universal evolution. Certainly, political institutions were one of the main subjects of research because it was through these that the basic distinction was made between "military societies" and "industrial societies."

Political theory thus appears to be the intellectual background against which we need to consider sociology's quest to establish a new science of society, or rather, the context within which the first attempts were made to develop the scientific subject matter on politics in the wake of the upheaval of the French Revolution.

4.1.3 *A new subject: the social question*

These were the beginnings of the era of sociology, but we must specify that its focal interest was the problem of the "social question" (the conditions of the lower classes) which had enormous importance in the process of formalizing sociology as an officially recognized discipline in the course of its historical development. Sociology addressed the conditions of these lower classes of society: poverty, unemployment, lack of housing and health care, criminality, and ethnic diversities. We need only recall the tradition of the English *social survey* with the well-known surveys on poverty by Henry Mayhew (1812 – 1887), Charles Booth (1840 – 1916), and Seebow Rowntree (1871 – 1954). Nevertheless, it was especially in the United States that *social work* research on the poverty of immigrant workers was a forerunner to this discipline, which would soon be studied in American universities, starting from Chicago.

The classic period of sociology coincided with the development of a critique of political economy and the attempt to face the problems posed by the social question.

Therefore, sociology arose as a movement renewing political theory under the impetus of the French revolution, and it was consolidated as a critique of the unrest caused by the industrial revolution. Reacting against the utilitarian-individualistic nature of free enterprise based on its principles of *laissez-faire*, the new social theories that developed in the last twenty-five years of the nineteenth century were inductive, socio-ethical, and interventionist. Sociology was part of this movement together with other new related disciplines, like historical economics in Germany and formal economics in the United States.

We can therefore identify three critiques with regard to political economy, each of which is of special importance in the development of the sociological project. One was centered on the politics of free enterprise.

The second, represented by the works of Max Weber and Emile Durkheim, highlighted the importance of the community founded on shared norms and values. The third laid the foundations for a critical analysis of the epistemological foundations of economics and the establishment of the scientific method of sociology.

4.2 *The discovery of society as a set of social relationships*

The birth of sociology can be considered to be the discovery of a new society founded on new social practices and relationships. The discovery of the existence and role of "civil society" emerged as a result. This "civil society" was the driving force of the upheaval resulting from the French Revolution. It was in France that sociology took form as a new theory that considered politics as a manifestation of broader and more general social processes.

Saint-Simon, Comte, and Tocqueville emphasized the fact that the frequency of political changes and the greater number of constitutions – solemnly proclaimed but short-lived – highlighted only superficially the legalistic value of political ideas. There was something new present in the social order, and this had to be discovered in the new social relationships and practices. In fact, not everything was negative, chaotic, and imposing terror and destruction. With the beginning of industrialization a new social force arose: the middle-classes.

The newly emerging sociology highlighted the contradictions between new social relationships and political forms, but it also sought to resolve the contradictions between the new industrial-democratic society and the anachronistic political set-up of the Restoration period. For Saint-Simon, Comte, and Spencer the new society was an "industrial society". For Saint-Simon the term had an anti-feudal connotation and intended to express a distinction between productive and unproductive labor:

> The crisis that has gripped the political system in the last thirty years can be traced essentially to the complete transformation of the social system. All the modifications that the old political arrangement has gradually undergone up to our day in the majority of civil nations stems from this transformation (Saint-Simon 1975: 182).

Comte's point of view was that the new society was characterized by a new economic activity organized in entrepreneurial form:

To clearly demonstrate the continual effect of industrial development on the general organization of the modern movement, I will first examine the influence of entrepreneurs and then of workers (Comte 1908, It. trans. 1967: 189).

In the writings of Spencer, an Englishman, we find an evolutionary outline based on a distinction between the internal and external processes of a system. In the case of social systems, Spencer indicates a contrast between economic activity and war. The two types of activity give rise to two different forms of social organization. Economic activity is voluntary and consists of the reciprocal interdependence of individuals who exchange services in a system of division of labor, whereas war is a coercive organization structured hierarchically and centralized (Spencer 1882: 189).

Therefore, in the England of Spencer's time as in the France of Comte and Saint-Simon's time, new economic activities shaped a society in which people did something very different from the past, a society in which productive activities replaced war as the dominant business of citizens.

In addition to these productive businesses, another interpretation of the new society stressed that there were new forms of social relationships among people. This was the idea of Tocqueville when, in his main work *Democracy in America*, he introduced the key concept of social democracy, referring to the conditions of equality and inequality existing in society.

The process of transformation that French society focused on was the egalitarian and democratic revaluation of the decline of the aristocracy and the rise of the middle class: «The gradual development of equal conditions is therefore a providential fact, and it has its own essential characteristics: it is universal, lasting and irreversible» (Tocqueville 1968: 189).

In the German experience, the fall of the old political order, after the French invasion, had highlighted another determining social factor in political institutions. This factor was not considered as something new but the rediscovery and reaffirmation of something old, such as the *volkgeist:* the "national culture". It was expressed in the language, ways, values, customs, and traditions of a nation.

German idealism then branched out into two currents: the Romantic current, represented especially by von Savigny's *historical school of law* and the Hegelian current. Although we do not find a conscious sociological tendency in German Romanticism, the sociological theory developed by Simmel, Tönnies, and Weber indicated that the origins of German sociology had its roots in Romanticism.

4.3 Society as a relationship: Tönnies, Durkheim, Weber, Marx, and Simmel

Thus the subject matter for classical sociological analyses is the discovery of new social relationships and behaviors that come with modernity.

In developing their scientific discovery, sociologists in the classical period worked to respond to the problems posed by the social question. They did so from a cultural perspective, criticizing the liberal viewpoint of political economics and the philosophy of natural law.

In Germany, this criticism and concern materialized in 1873 with the founding of the journal *Verein Für Sozialpolitik*, which involved all the major German sociologists and economists who advocated for "ethical economics." One of these was Gustav Schmoller (1838-1917) who supported the idea of inductive and empirical economics: economies and capital did not exist in the real world and could not, therefore, be considered as distinct phenomena, in isolation from the contexts in which they were operative.

The *Verein* had considerable influence on American sociologists, especially those who belonged to the *American Sociological Society*. One of these, Albion Small, an ardent admirer, dedicated special attention to it in his work on the origins of sociology in America (Small 1911). However, the sociologist who adopted the cultural heritage of *Verein* from overseas was a young professor who had studied Sombart and Weber while in Germany. His name was Talcott Parsons (1902-1979). His studies of the German economic tradition led him to write critiques on classical political economics and socialism, both of which were imbued with utilitarianism. He did so in order to attribute a decisive role to the norms and values shared in social relationships. His conclusions constitute the true subject matter of his work *The Structure of Social Action* (Parsons 1937).

The first great classical work dealing with social relationships is that of Ferdinand Tönnies (1855-1936) in his *Gemeinschaft und Gesellschaft* – Community and Society (1887). The warmth and harmony of the family community and the village were exalted and set against the calculating and pragmatic egotism of society. More generally, Tönnies intended to present a theory of "human wills" based on reciprocity and the rapport between the unity and plurality of human associations. He wrote:

> The present theory will exclusively assume the relationship of reciprocal affirmation as the subject-matter of their own surveys. Each one of these rapports represents a unity in plurality or a plurality in unity.... The group formed by this positive rapport, conceived of a subject acting in a unitary

manner internally and externally, can be called association. The relationship in itself and therefore, the *association*, is conceived either as real and organic life – and this is the essence of the *community* – or as ideal and mechanical formation, and this is the concept of *society* (Tönnies 1887, repr. 1963: 45).

Tönnies described the embryonic forms of community distinguishing three kinds of relationships: 1) the relationship between mother and child; 2) the relationship between husband and wife; and 3) the relationship among those who recognize themselves as brothers and sisters. Tönnies emphasized that sibling love is the most human relationship that can exist and the most authentically communitarian. Furthermore, he pointed out that this relationship is based on love and reciprocal will. He said:

> Sibling love can be considered as the most human relationship among human beings, even though it is still completely based on blood relations. We see this in effect where instinct is weakened by all the causes of hostility which could negatively affect this relationship. Memory seems to cooperate to maintain and strengthen the bonds of the heart by recalling all the pleasing impressions and experiences associated with the person and his or her actions (Ibid: 52).

In social life, however, Tönnies identified only thirst for power and money in individuals who build merely instrumental relationships. With regard to relationships in society, he said: «Personal interests and vanity are the motives of sociability; vanity needs other people as mirrors, personal interests need them as instruments» (Ibid: 458).

What emerges from his analysis is that social relationships are the foundation of collective living. If on one hand they have a classifying importance, from the empirical viewpoint they seem to be rather limited. Nonetheless, in accord with the analyses of the previous authors, Tönnies emphasized a process of radical change such as the move to modernity in which different relationships characterize different types of society.

Durkheim's main sociological work *The Division of Labor in Society (1893)* in many respects also constitutes a work on social relationships and their transformation in the modern era. Durkheim wrote many texts of a sociological nature. We may mention among others: *The Rules of the Sociological Method (1895), Suicide (1897), and The Elementary Forms of Religious Life (1912).* It is to him that we owe the development of sociology, and he probably gave the greatest contribution ever to making it a subject of study, besides developing sociological theory.

Durkheim began with an observation: «A totality is not identical to the sum of its parts. Rather, it forms something else, whose properties differ from those present in its composing parts». It is from this supposition that the identification of the specific subject-matter of sociology ensues: «Association... constitutes the source of all novelties.... In virtue of this principle, society is not simply a sum of individuals... but a specific reality endowed with its own characteristics» (Durkheim 1895, It. trans. 2001: 56-57). In discovering the existence of society, Durkheim defined one of the basic theoretical problems of sociology still prevalent today: the relationship between two given entities – the individual and society. This was the moral, sociological, and political question of Durkheim's entire work. The moral question consisted in how to harmonize individual freedom and social order. The sociological question consisted in demonstrating the existence of society insofar as it is a distinct reality from its individual components. The political question was how to ensure both individual freedom and collective solidarity.

Durkheim was one of those sociologists who focused on social relationships which give life to social events and social integration, two of the most important sociological concepts developed by Durkheim. According to him, interaction between individuals constitutes a reality which cannot be explained by biological or psychological factors. Consequently, he shows that relationships between individuals give life to "social events".

Durkheim is remembered for being the founding father of a theoretical paradigm called Functionalism. Some of the major contemporary sociologists, like Parsons, Merton (1910-2003), and Luhmann (1927-1998), can be counted among his followers.

Another outstanding sociologist was Max Weber. His vision was a long way from that of the founding fathers of functionalism – a natural science of the evolution of humanity and society guided by the rise of the middle-classes. At the same time, however, he opposed both the Hegelian and Romantic idealistic schools of thought.

The principal category in Weber's analysis and in the methodological program of the new science was social action characterized by meaningful relationships between two or more subjects. In his writings we read: «By 'social' action, we must intend an action that is directed towards other individuals and which is consequently conditioned by them» (Weber 1922, It. trans. 1961: 63). Society is born from two or more subjects who have a mutual and meaningful exchange. This meaningful dimension in a social interaction is the foundation of a new sociological methodology based on the understanding of the meaning in this exchange, constituting a social relation.

Therefore, Weber was the first sociologist to define social relationships as being based on the reciprocity of action and meaning:

> By social relationships we must intend a behavior of reciprocity between several individuals that is meaningful and consistent. Consequently, the interaction must be characterized by reciprocal actions on the part of both (Ibid: 78).

Another characteristic of society is the reciprocity of people's actions and their openness to others. Of course, nothing is said about the nature of this relationship which can vary from one of gratuitous giving towards a neighbor to open hostility, conflict, and the exclusion of others.

For Weber, the meaning that the subject attributes to his or her action is paramount to the concept of social relationships. Thus we have societies where a meaningful subject interacts with the behavior of other individuals and is consequently conditioned by it. In this perspective, also for the so-called "social institutions" – like the State, the Church, marriage, and so forth – social relationships consist exclusively and simply in the possibility that an action reciprocally established took place, is taking place or will take place in a given way, according to its meaning" (Ibid: 79).

Thus Weber introduced interaction between subjects as the starting point for great historical events and macro social formation for the first time in Western thought. The topics on which this great intellectual reflected – the State, modernity, capitalism, bureaucracy, power, cities, and religion – indicate some of the categories which comprise meaningful actions between interacting subjects.

Weber is remembered for his methodological writings and for having developed an analytical tool which is ideal in an investigative methodology that analyzes, understands, and interprets actions.

Perhaps neither Durkheim nor Weber would be who they are for us today had they not known the intellectual work of Karl Marx (1818-1883).

As is known, Marx was a political theorist, economist, philosopher, political ideologist and leader, and also a sociologist. It was Marx who defined modern society as "capitalistic", a notion which was implanted in the reflections of Durkheim and Weber. We owe to Marx the empirical intuition of the relationship between social structures and ideas, as well as the concept of social classes. He was the one who developed the complex theoretical model of conflict. It is not my intention to give a systematic presentation of Marxist theory, but it is relevant from the viewpoint of social relationships.

In a capitalistic society, social relationships are in constant flux and it is through these conflicting tensions that social change is generated. In the first place, it must be said that Marx saw historical subjects in a collective manner. Indeed, he had a holistic approach to society. From an analytical viewpoint, there are only "productive forces", that is, people establish relationships with one another in the incessant struggle to snatch from nature their means of a livelihood. This is the driving force of history: «The first historic action is... the production of material livelihood» (Marx 1972, vol. V: 27).

"Productive forces" enter into "relationships of production." With this concept Marx intended to highlight all those social relationships which people establish through their participation in economic life. Therefore, the relationships of production are not only machines which produce, but also the production processes and the organizational aspects of production.

These "relationships of production" create collective subjects which are the social classes. In the preface to *Capital*, Marx states his methodological premise defining the subject and type of his analysis: «We are dealing with people here only inasmuch as they personify economic categories; they represent certain relationships and class interests» (Ibid, vol. I: 18).

In fact, Marx did not objectify society or classes; he recognized a degree of autonomy in the subject. Nevertheless, his thought is negatively influenced by naturalistic and mechanistic epistemology which always compels him to indicate one factor that, in the final analysis, determines the others.

However, from our viewpoint of "social interaction" Marx also introduced the concept of "alienation" to the analysis of the relationships that are established in a capitalistic society, a topic particularly important to the German cultural tradition. Marx held that all social institutions tend to alienate inasmuch as individuals lose the awareness that they themselves are the authors and builders of those very institutions. They do not see their actions connected with those institutions.

This process is characteristic of the working world and is expressed in four aspects of alienation: a) from the objects they produce, b) from the process of production, c) from oneself, and d) from one's community:

> the estrangement of one person from another... to say that a person's very essence is estranged from another means that one person is estranged from another, as each one of them is estranged from their human essence (Ibid, vol. III: 304).

Relationships of exploitation, alienation and conflict are a heritage of sociological thought, thanks to the Marxist analysis.

Another great contribution to sociological thought was made by a German sociologist: Georg Simmel. Simmel can be defined as the sociologist of interaction. Indeed, he is the thinker that people refer to when considering symbolic interaction.

In his writings we find a definition of society based on the reciprocity of the actions of individuals: «Society exists wherever several individuals enter into reciprocal actions. Such actions are the result of specific impulses or in view of specific goals».

For Simmel, the reciprocal action of separate individuals does not build a society unless unity emerges between the parts: «These reciprocal actions mean that a unity arises between the individual bearers of those occasional impulses and goals, that is, a 'society.' In fact, unity in the empirical sense is the result of the reciprocal action of elements».

Subsequently, Simmel stated that society comprises the unity of reciprocal actions, referring to the actions of daily life – those infinitely numerous and infinitely small actions:

> that unity or association can be present in very different degrees, depending on the mode and proximity of the reciprocal action – from short-lived gatherings such as a family gathering, or all valid connections, even retracting one's citizenship, to the passing encounters between acquaintances in a hotel, to the close connections in a medieval guild (Simmel 1908, It. trans. 1989: 105-106).

The social order founded on daily interactions is the subject of Simmel's analysis:

> only what takes place in the domain of physical and spiritual contacts, of mutual actions which give rise to pleasure and suffering, of conversation and silence, of common and antagonistic interests – this alone constitutes the wonderful indissolubility of society (Ibid: 119).

However, Simmel is also the sociologist who devoted himself to developing a "sociology of interiority". In 1907 he wrote an essay entitled "Gratitude" (Simmel 1996: 91-103). This sentiment became for Simmel one of the strongest cohesive forces of society and if it were lacking, society would disintegrate, at least as we know it. Gratitude is the link that keeps us united, a link that is inadequate in expressing our thank you for a gift received from someone who "acted first" in total freedom and gratuitously gave to us.

Conclusions

The discovery of society and new sociological reflection begin with the observation of new practices and social interactions among individuals in the new era of modernity, and urban and industrial society. Social interaction is the element which distinguishes the sociological realm from the reflection on the social sphere that preceded it. This new theoretical reflection arose in a politically, economically, and scientifically revolutionary historical context. The object of sociological thought comes from conflict with other forms of knowledge. From this point of view, cultural conflicts are an expression of social struggle and power (Lepenies 1987).

The scientific knowledge of social action was caused by the revolutions of the eighteenth century, with the conflict between theological and the scientific thought. Instead, the first great sociological theories matured between the nineteenth and early twentieth centuries, fighting against the utopian vision, and ideological and authoritarian regimes.

Bibliographical references

Comte, A., *Cours de philosophie positive* (Paris, 1908), IV; It. trans. *Corso di filosofia positiva* (Torino: Utet, 1967), vol. II.

Durkheim, E., *Les règles de la méthode sociologique* (Paris: Félix Alcan, 1895); It. trans. *Le regole del metodo sociologico. Sociologia e Filosofia* (Torino: Edizioni di Comunità, 2001) in A. R. Calabrò, *Oggetto e metodo della sociologia: parlano i classici* (Napoli: Liguori, 2003).

Lepenies, W., *Le tre culture* (Bologna: Il Mulino, 1987).

Marx, K, 'L'ideologia tedesca, Critica della più recente filosofia tedesca nei suoi rappresentanti Feuerbach, B. Bauer, e Stirner, e del socialismo tedesco nei suoi vari profeti', in K. Marx and F. Engels, *Opere* (Roma: Editori Riuniti, 1972), vol. I.

— 'L'ideologia tedesca, Critica della più recente filosofia tedesca nei suoi rappresentanti Feuerbach, B. Bauer, e Stirner, e del socialismo tedesco nei suoi vari profeti', in K. Marx and F. Engels, *Opere* (Roma: Editori Riuniti, 1972), vol. III.

— 'L'ideologia tedesca, Critica della più recente filosofia tedesca nei suoi rappresentanti Feuerbach, B. Bauer, e Stirner, e del socialismo tedesco nei suoi vari profeti', in K. Marx and F. Engels, *Opere* (Roma: Editori Riuniti, 1972), vol. V.

Nisbet, R., *Storia e cambiamento sociale. Il concetto di sviluppo nella tradizione occidentale* (Milano: Isedi, 1977).

Parsons, T., *The structure of social action* (New York: McGraw-Hill, 1937); It. trans. *La struttura dell'azione sociale* (Bologna: Il Mulino, 1962).

Saint-Simon, H., *Opere* (Torino: Einaudi, 1975).

Simmel, G., *La gratitudine*, in G. Simmel, *Sull'intimità,* ed. by V. Cotesta (Roma: Armando Editore, 1996).

— *Soziologie* (Leipzig: Duncker & Humblot, 1908); It. trans. *Sociologia,* (Milano: Edizioni di Comunità, 1989) in A. R. Calabrò, *Oggetto e metodo della sociologia: parlano i classici* (Napoli: Liguori, 2003).

Small, A., *Origins of Sociology* (Chicago: University Chicago Press, 1911).

Spencer, H., *The Study of Sociology* (London: Henry S. King and Co, 1882).

Tocqueville, A., *La democrazia in America* (Torino: Einaudi, 1968).

Tönnies, F., *Gemeinschaft und Gesellschaft* (Lipsia: Verlag di Fues, 1887); It. trans. *Comunità e società* (Milano: Edizioni di Comunità, 1963).

Weber, M., *Wirtschaft und Gesellschaft* (Tübingen: Mohr, 1922); It. trans. *Economia e società* (Edizioni di Comunità, Milano, 1961) in A. R. Calabrò, *Oggetto e metodo della sociologia: parlano i classici* (Napoli: Liguori, 2003).

GIORGIA MAVICA, DAVIDE NICOLOSI, ALESSANDRA SCIERI[*]

5.
THE RELATIONSHIP BETWEEN SOCIOLOGY AND HISTORY IN THE CLASSICS OF SOCIOLOGY

Although they have in common several objects of investigation, sociologists and historians have rarely shown collaboration in interdisciplinary research or use of one another's expertise to achieve more effective and extensive results. In other words, they look at reality in a different way: history in particular, by dealing with the uniqueness of events, aims to understand facts that happened only once and is less interested in the regularity with which phenomena occur. On the contrary, as a science oriented to the generic, sociology aims to study the historical development of societies and the social actions of men with regard to their typical features, i.e., the characteristics that recur in several cases. If sociology was born to study social change, it is bound to have close links with history, the objective of which is to reconstruct social events within evolutionary processes.

Already at the beginning of the last century Weber recognized the indissoluble link between sociology and history, based on a mutual and indispensable support and on a logical precedent, according to which a kind of sociology that is based on the present and a kind of history that operates only in the short term risk losing the historical dimension of time and not capturing the scope of long-term processes and inevitably turning into a sociology and a history of the contingent (Cavalli 1963: 220). Hence the need to reconstruct social processes within a relationship in which time represents the foundation of the relationship between sociology and history and in which there is a narrative continuity between past, present and future in order to understand human action.

Sociologists analyze social structures through a static approach, while historians, on the other hand, thanks to a dynamic approach, investigate above all the changes that have occurred in social structures. However, if these continually undergo social mutations and each of them affects the structures

---

[*] Although this chapter is the result of a common reflection among the authors, Giorgia Mavica wrote the second paragraph, Davide Nicolosi wrote the first and Alessandra Scieri the third.

of society, both fields of study should therefore integrate better, in order to achieve a more solid integration between the two disciplines which could, in this way, be defined "historical sociology" and "social history" (Burke 1982: 7). «The separation between sociology and history is disastrous and totally without epistemological justification: every sociology must be historical and every history sociological» (Bourdieu 1992, It. trans. 1992: 62).

The sociological tradition is characterized by the presence of a fundamental dilemma that is revealed through the question that lies at the basis of the whole of sociology, that is to understand if it is the individual who creates the context or vice versa, and the extent to which man is free to act or is conditioned by some external entity. It is from here that the various paradigmatic antinomies take shape, such as holism and individualism, structure and individualism, and society and action present in the thought of the great classical theorists. Starting from the above mentioned antinomies, the aim of this chapter is to analyze the relationship between history and sociology and the different connotations that emerge in the thought of three great authors: Marx, Simmel, and Weber.

The purpose will be to outline their thought regarding the possible relationship between the two disciplines, the way in which – albeit starting from different epistemological positions – they decline logical structures and perspectives of analysis that take into account the following condition:

> There is a very close solidarity between history and sociology, between the analysis of events and the establishment of general propositions. The understanding of the social fact requires the use of general propositions, and these can only be demonstrated on the basis of analysis and historical comparisons (Busino 1975: 65).

5.1 *Karl Marx starting from historical materialism*

Marx saw society as an unstable balance of opposing forces, which, through their struggles and tensions, generate social change. Consequently, he envisaged the development of society in an evolutionary sense, considering social conflict as the determining element of the historical process.

In this regard, starting from historical materialism it can be seen how the author intended to study society in an evolutionary sense, conceiving the aforementioned materialism as a way of thinking, starting from the analysis of the material conditions of humans, as they have been historically determined. In particular, historical materialism is a set of criteria and theoretical assumptions with which Marx analyzed and interpreted the history of humans in different societies, particularly in capitalist societies (Cerroni 1976: 13). The history of humans is understood as a history of the ways in which they have organized themselves together to produce, that is, to relate to resources and therefore to nature in order to ensure their survival. Within this context, "the division of labor" plays an important role because the ways in which labor is divided, together with how ownership is distributed and the production techniques that are historically determined from time to time, form the material basis of a society – what Marx calls "structure". In Marx's thought, the idea of "structure" is fundamental because it determines the forms of the "superstructure", understood by the author as the areas of legal institutions, religious representations, morals, and philosophy: this means that these forms do not have a history of their own, but depend on their unfolding and the modifications of the structure to which they correspond (Del Vecchio 1964: 5). However, unlike other structuralist sociologists, Marx observed that the "division of labor" always seems to have been unequal, developing another crucial concept in his thinking – that of ideology, considered as a set of propositions that represent the world in a partially falsified way: that is, a representation of the world that describes and, at the same time, conceals its real conditions, in particular the contradictions and conflicts present within society, tending to substantially immobilize history (Jedlowski 2009: 46).

Although Marx followed Hegel's dialectical model, according to which history evolves through constant contradictions and conflicts, he differed from Hegel in that he saw dialectics as an active principle operating within material conditions and objective economic and social relationships. In fact, the Erfurt scholar maintained that historical and social evolution is determined by objective contradictions linked to the availability of "material and technical resources" and "production relationships", which from time to time are established in different historical periods until an egalitarian and perfectly integrated society is reached through the elimination of private property and the organization of the relationships connected to it (Crespi 2002: 38).

Marx believed that the Hegelian idealism must be overcome because Hegel's perspective prevents us from understanding humans as "real beings", simply reducing them to pure abstraction. This aspect becomes the pivotal point for a real and positive science of society for Marx, who also points out that rather than considering people as "abstract" and "isolated" individuals, one should consider them as individuals whose human essence is actually provided by all social relationships (Ghisleni 1998: 83-84).

For Marx, the isolated individual is inconceivable, or rather, he or she is conceivable only under certain historical conditions: for example, modern society is a society in which the division of social labor is highly developed and the form in which the products of this divided labor come together is the market. However, the market is a system of "abstract" relationships, i.e., in this case, individuals do not exchange their products with regard to personal relationships but on the basis of impersonal laws, represented by the price of goods: however, with regard to the market, each individual is effectively isolated, since his or her relationships with others pass through the goods, which seem to be subject to the laws of the market itself. Moreover, the division of labor that developed with the capitalist mode of production is so complex that it leads each individual to confine himself within a certain role: in this "confinement", an aspect of people's alienation is hidden because they work and produce in such high proportions (never imagined in past historical times) that they cannot enjoy relationships with other people. Therefore, for Marx, a human being is a "social being" who needs to live with his or her fellow human beings, and his or her own human consciousness is produced by social interaction (Jedlowski 2009: 58).

Marx does not consider the history of capitalist society to be the only one in which two antagonistic classes have opposed each other, but for the scholar the history of every society that has existed up to that moment is actually the history of class struggles, as Engels argues, taking up the thought of his colleague:

> [...] that economic production, and the structure of society of every historical epoch necessarily arising therefrom, constitute the foundation for the political and intellectual history of that epoch; that consequently (ever since the dissolution of the primaeval communal ownership of land) all history has been a history of class struggles, of struggles between exploited and exploiting, between dominated and dominating classes at various stages of social evolution [...] (Marx and Engels 1848, Eng. trans. 1985: 57).

Through an analysis of human history in different historical periods, in particular in the study of the division of labor in capitalist society, one can identify an important step that allows us to see the relationship between history and sociology as they are linked in Marx's thought. First of all, the first premise of all human history is essentially to consider humans as living beings who, in order to ensure their very survival in the natural context, need material organization through their interaction and transforming action, thus producing their living conditions and means of subsistence. This means that the way individuals organize themselves depends on the material conditions of their production. This appears with the increase in population, particularly when the different forms of ownership came into being, where it is already possible to trace the first divisions of labor. In particular, the social classes began to develop when men who occupied certain social positions or roles found themselves involved in particular social circumstances: for example, in the societies of Ancient Greece and Rome the economy was organized around slavery, so that the main classes were formed by those (the patricians) whose livelihoods came from slaves and the slaves themselves (the plebeians); in the feudal society of the Middle Ages the basis of the economy was the fiefdom; the main class was the nobility, who owned the land and the services of the peasants tied to it, and the serfs or the peasants who worked the land itself (Marx and Engels 1949: 9-13); on the other hand, within capitalist societies competition divided the personal interests of the proletarians, but at the same time, the establishment of the wage system led to the formation of two classes: the class of the capitalists and that of the proletarians – two classes whose relationship is essentially based on those who take control of the labor force (the capitalists) and those who are forced to sell it (the proletarians) (Coser 1997: 87). For Marx, the conflict between the antagonistic classes derives mainly from the "value-work" ratio whereby value refers to the amount of work needed to produce it. The "value-work" theory undoubtedly attracted a socialist like Marx, because if work produces value, then workers expect to receive the proceeds of their work. In spite of this, through observations and studies that he made on the capitalist system, Marx maintains that when faced with this theory, we are actually facing a paradox: an entrepreneur may decide to pay a worker for six hours, but make him work for eight. This is because the entrepreneur is the owner of the means of production, and therefore owner of the factory and the tools, and can force him, as a condition for giving him a job, to do a longer day's work. Therefore, the profit comes from the extra working hours, over and above those paid to the worker. This extra work is called "surplus value" and so it can be said that profit is essentially based on the exploitation of labor (Collins and Makowsky 1980: 39-40).

The exploitation of labor is closely linked to Marx's concept of "alienation" and should be understood with reference to his theory of value in which he considered productive activity as the essence of man. The author considers alienation as that specific situation in which the worker does not have the possibility to reap the benefits of his work, as he is forced to sell his own activity, which is his own humanity, to others, thus finding himself extraneous to the result of his work. In this way, the worker is not only alienated from the finished product of his work, but from his own work activity, which essentially does not belong to him. To be estranged from one's work activity means to lose one's "essence" as a person; it means alienation from oneself and from one's humanity. There are many philosophical and youthful writings by Marx that deal with this subject:

> if the product of labor does not belong to the worker, if it confronts him as an alien power, then this can only be because it belongs to some other man than the worker. If the worker's activity is a torment to him, to another it must be delight and his life's joy (Marx 1844a, Eng. trans. 2012: 79).

The situation of alienation in which man finds himself according to Marx is the result of the capitalist economy, based on the substitution of the value of use, linked to the actual needs of the individual, with the value of exchange, considering the individual himself as purely a means of exchange or a commodity. Alienation is therefore considered by the scholar as a purely internal phenomenon of thought, the result not only of psychological conditions, but also of historically determined economic and social structures. According to the author, in order to overcome alienation it is necessary to eliminate the economy of exchange itself, to give back to the worker the direct control of the means of production and finally, to re-establish a correct relationship between productive activity and actual human needs (Marx 1857-58, It. trans. 2010: 49).

In the light of what has been said so far, it is possible to infer the link between sociology and history in the author's thought: people are not only social beings, but they are also historical beings and, at the same time, they are determined by history itself because as well as being defined by their relationships with others, their identity is determined by the historical process, since the institution of private property led proletarians to alienate themselves at work (Izzo 1991: 104).

Therefore, according to Marx, the history of mankind has a twofold aspect: it is both the history of mankind that gained increasing control over nature and the history of his progressive alienation. The latter can be defined as a condition in which people are dominated by forces which they themselves have created and which oppose each other as alienated forces (Coser 1997: 90-91).

In this case, the proletarian class, to define it in the full sense, should be bound to class consciousness and this would allow the passage from "class within itself" to "class for itself". According to Marx, class consciousness depends on the material resources available to produce and exchange ideas, and it is basically the dominant social classes that possess most of these resources. As history progresses, class awareness has also changed: in fact, Marx believed that during the Age of Enlightenment the nascent bourgeois class began to have sufficient resources, such as the spread of elementary education and the press, which allowed workers to gather in factories and communicate with each other and begin to develop their own worldview. Thus, alongside the official ideology of the bourgeoisie, a new underground consciousness was beginning to take shape (Collins and Makowsky 1980: 35) on the part of the proletarian class – a consciousness in which man conceives self-fulfillment as a result of the commitment he puts into his work, hence into the object of his production. Work activity, understood as the final phase of the realization of the human essence, must not become the "means" of life, but life itself. Consequently, Marx maintains that only through the suppression of private property can man acquire his essence and his freedom.

In *The Jewish Question* Marx takes into consideration the bourgeois Constitution, in particular the French Constitution of 1793, and highlights the concept of freedom, which is essentially based on the right to private property, which allows men to think about their own interests without having any regard for others or society itself. The limit for man to acquire his own essence is inherent within this freedom, and this limit highlights the fact that only a part of civil society had emancipated itself and had really come to the possession of power; only a part had undertaken the emancipation of society from its own particular conditions. For Marx there exists the possibility of a deeper and more widespread emancipation, given by a class that is the dissolution of all other classes, a "universal" class that has no particular rights to claim, that does not claim historical titles, but only human titles because it is in universal contrast with all the premises of the social political system (Costantino and Zanca 2015: 48-49). This class is the proletariat, as the scholar of materialism says: «a class with radical chains» (Marx 1844b, Eng. trans. 1970: 141).

In the *Manuscripts*, Marx outlines a hypothesis of communism that takes into account the variety of human needs and is based on a new mode of production capable of realizing the human essence and promoting its enrichment. Taking this into account, according to the scholar, communism

> is the true solution of the struggle between existence and essence, between objectification and self-affirmation, between freedom and necessity, between individual and species. It is the solution to the riddle of history and knows itself to be this solution (Marx 1844a, Eng. trans. 2000: 97).

Therefore, according to Marx, the transformation of bourgeois society into a classless one is the responsibility of the proletariat, since it is not the bearer of special interests that are opposed to each other, but rather the bearer of the interests of the whole of humanity. This should, finally, lead to a classless society, to the kingdom of freedom, and the possibility of realizing oneself freely, leaving others free from those coercive, economic and political situations, which had characterized all the previous history of humanity (Izzo 1991: 113). This previous history is actually "prehistory", as a kingdom of necessity, while for Marx true history would begin with the state of freedom:

> the bourgeois relations of production are the last antagonistic form of the social process of production – antagonistic not in the sense of individual antagonism, but of one arising from conditions surrounding the life of individuals in society; at the same time, the productive forces developing in the womb of bourgeois society create the material conditions for the solution of that antagonism. This social formation constitutes, therefore, the closing chapter of the prehistoric state of human society (Marx 1859, Eng. trans. 1904: 13).

Therefore, from Marx's studies it is possible to gather the elements of a method that led him to lay the foundations for a new way of understanding the social sciences that was increasingly linked to history. Indeed, taking into account what has been said so far about Marx's thought, we can focus on three important points that allow us to identify an ambivalent link between history and sociology. First of all, through his existence man creates history and this implies material acts that allow him to satisfy his needs such as eating, drinking, having a roof over his head, clothing himself, and so on. When he produces the means to satisfy these needs, all he does is produce material life and therefore history. Secondly, another need that man needs to satisfy is to enter into interaction with the other, understood as a natural and social act, which develops from his consciousness. Initially it was a question of awareness of existence like that of animals, and only later with development and with the increase in productivity, with the increase in needs and population did man develop a human consciousness determined by the relationship and interaction with his fellow human beings, with whom he created the history of societies based on the division of labor. In the third and last point, Marx highlights the passage from the division of labor to the development of social power: «the multiplied productive force, which arise through the cooperation of different individuals» (Marx and Engels 1845-46, Eng. trans. 1978: 161).

Social power appears as a power extraneous to the individuals themselves, who are pushed towards a state of alienation, which, according to Marx, can be overcome by acting against power through the establishment of communism, that is, a classless society understood as the "kingdom of freedom". The last point highlighted in Marx's thought states that it is not only society that determines history, but it is possible to think inversely about this bond, since the action of individuals within certain classes creates history, but also the very history of individuals can determine their social conditions through an awareness of the state of alienation in which they find themselves, as in the case of the division of labor in capitalist societies. Indeed, in agreement with Engels, Marx states that: «it is not consciousness that determines life, but life that determines consciousness» (Marx and Engels 1845-46, Eng. trans. 1976: 37).

Moreover, in the studies of Marx's works one can not only learn this ambivalent link between history and sociology, but also that the former is enhanced by the latter precisely because the study of the history of societies can be understood in its entirety starting from an empirical investigation of concrete social processes, which is a condition *sine qua non* of human existence (Giddens 2009: 57). Consequently, only in this way does "history cease to be a collection of dead facts":

> this method of approach is not devoid of premises. It starts out from the real premises and does not abandon them for a moment. Its premises are men, not in any fantastic isolation and rigidity, but in their actual, empirically perceptible process of development under definite conditions. As soon as this active life-process is described, history ceases to be a collection of dead facts as it is with the empiricists (themselves still abstract), or an imagined activity of imagined subjects, as with the idealists (Marx and Engels 1845-46, Eng. trans. 1970: 47-48).

Regarding Marx and his relationship to history, Mills (1974: 198) argues the importance and value of history in understanding the development of societies over the course of historical time by saying that we cannot understand any single society, even as a static fact, if we do not use historical material. In fact, this scholar takes up the Marxist concept of the "principle of historical specificity", that is, the image of any society is a specifically historical image: in this case, according to Mills, no given society can be understood except in terms of the specific period in which it exists. In whichever way one may understand the word "period", the institutions, ideologies, human, male and female types, predominant in a given period, represent unique examples. This does not mean that that exemplary historical data cannot be compared with others; on the contrary, through comparative studies between societies in different historical periods we can become aware of the fact that some historical phases are absent from a particular society we are studying: for example, taking into account what was said earlier about the analysis of societies of the past, Marx believed that unlike the proletarian class, the slaves of Ancient Greece and Roman society did not have the possibility to found a society like the communist one because they did not have sufficient educational and material resources to enable them to unite and develop their own worldview, and therefore to found a classless society, like the one understood by the scholar of historical materialism.

Therefore, if we put history aside, it is impossible to understand precisely the current features of the society we want to study or analyze; we cannot hope to understand the structure of a society if we are not guided by the sociological principle of historical specificity (Abd el-Malek 1974: 90-91).

5.2 *"Historical knowledge" in Georg Simmel*

When carrying out research in sociology, one thinks of the classics not only as pioneers of sociological thought, but also as essential frames of reference. In this sense, Georg Simmel has undoubtedly provided important input for sociological research and has been a source of inspiration for entire generations of scholars. Simmel, not unlike Marx and Weber, avoids describing society through reference to the metaphor of the living organism; in fact, he considers it a purely ideal term. Society is not "tangible in any given place", which is why sociology must be oriented towards the analysis of associations. These are the only aspects of empirically observable reality (Simmel 1918, It. trans. 1976: 55).

Simmel's sociological approach can rightly be considered as an attempt to contrast both Comte's and Spencer's theories of the organic, and the thesis – then in vogue in Germany – of the historical description of the phenomena considered in their individuality. Simmel maintains that society is made up of a set of formal interactions and that the task of the sociologist is to study the forms of such interactions as they occur in cultural situations in different historical periods (Coser 1983: 257).

It seems clear that while following his own original approach, Simmel re-proposes the concept of the historical liberal model found in such different thinkers as Spencer and Durkheim. In relation to this, Simmel's concept of differentiation implies the passage from homogeneity to heterogeneity, from uniformity to individualization, and from assimilation through repetitive activities, typical of a small world based on tradition, to participation in a wider world based on multifaceted commitments. In this perspective, Simmel emphasizes the existence of an inevitable dualism inherent in the relationship between individuals and objective cultural values; the individual is continually confronted with a multiplicity of cultural contents, such as religion, morality, custom, and science. In order to achieve autonomy and achieve their goals, individuals need to internalize these cultural values by making them part of themselves (Ibid: 233-234).

Culture also comes to be constituted as an autonomous complex with its own objective spirit that contrasts with the subjective spirit, i.e. individual: the individual cannot dominate the enormous cosmos constituted by objective culture that arises as an autonomous entity (Izzo 1991: 166).

Simmel's sociology goes far beyond the salons, tackling problems related to power, money and historical change with profound insight (Collins and Makowsky 1980: 161):

> Simmel was more an activist than a systematic person, more someone who made philosophical pronouncements about his time with a scientific-social touch than a well- established philosopher or sociologist rooted in scientific activity (Schnabel 1976: 16).

Already in his first works, in a field of investigation ranging from history, anthropology, and philosophy to law and psychology, Simmel carried out his research starting from the assumption that the object of investigation of the human sciences (*Geisteswissenschaften* or *Kulturwissenschaften*) cannot be considered as a simple "fact". Although the influences of the young Simmel reveal an "evolutionist" component (Meschiari 1996), his interests – and possible applications and implications in the field of humanities – shift towards the "neo-Kantian" investigations of his esteemed colleagues Wilhelm Windelband and Heinrich Rickert and towards those of a "historicist" nature of his colleague Wilhelm Dilthey. (Magnano San Lio 2005: 45). One of Simmel's first works offers the possibility of analyzing the question of "historical knowledge" in relation to reflections and debates on this subject: in the text *Die Probleme der Geschichtsphilosophie* (1892) one can therefore understand the desire to create a theory of historical knowledge, whereby "history" for Simmel means a theoretical construct that is the result of the methodological convergence of psychology and logic: psychology studies the "processes of psychic realization", while logic, as a science of objective conceptuality, deals with the "essential contents" (Simmel 1908, It. trans. 1989: III):

> History is somehow something intermediate between logic, the purely objective consideration of our psychic contents and psychology, the purely dynamic

consideration of the psychic movements of these contents (Simmel 1890, It. trans. 1982: 8).

The probability of reconstructing a historical fact is determined by the *a priori* possibility of "understanding" the studied phenomenon: Simmel's work is an almost anthropological construction of the theory of historical knowledge. This sort of "phenomenological reduction" of the emotional-psychological life of the human spirit inevitably refers to the revolution accomplished shortly afterwards by Max Scheler (D'Anna 2006: 25).

History is never the homogeneous copy of reality or of real events: rather, it is the result of a hermeneutical work that is placed on a plane of intersection thanks to a "relational logic" (Simmel 1890, It. trans. 1982: 33). «Historiography is presented as the observation of external data and their interpretation, as a description of singular phenomena and as a conceptual synthesis» (Simmel 1892, It. trans. 1982: 118). History is attentive to the collection of "objective" data, but Simmel denies the possibility that the historical fact corresponds to the simple mechanical sum of these data: history must aim at understanding the development of psychological contents and not at their simple collection. On the other hand, the same equivocal concept of objectivity must be clarified: «the value of objectivity [in the field of historical knowledge] is not linked to the autonomous coherence of the object» (Simmel 1890, It. trans. 1982: 58). According to a specifically (neo) Kantian methodology, Simmel asserts that the possibility of reconstructing a historical fact is given by the ability *a priori* to "understand" the studied phenomenon, immersing oneself in it: Simmel's work is an almost anthropological foundation of the theory of historical knowledge. The spiritual nature of man is based on "feeling" and "will": representation in the field of human sciences is conditioned by these *a priori* fundamentals.

> If the task of history is to know not only the object of knowledge, but also of will and feeling, this task can be fulfilled only by participating, in some form of psychic transposition, with the will to what is willed and with feeling to what is felt (Ibid: 32).

The social-historical object must be investigated in a relational way in an "intersubjective" dimension. No historical object exists without a dimension of openness towards otherness being presupposed at the origin.

> If one considers history as objectivity and not as a function of the thinking spirit, the problems of its philosophical consideration emerge, that is, as far

as I can see, two distinct groups of particularly characterized problems. The first is based on the fact that history is a sum of empirical details. The question therefore arises, on the one hand, whether the whole has a strange nature and meaning that cannot be inferred from any particular element as such, and on the other hand – but perhaps coinciding with the first – the question of which absolute being, which transcendent reality is behind the phenomenal character of empirical-historical data (Simmel 1892, It. trans. 1982: 118).

At the basis of Simmel's sociological analyses there are almost always questions or issues that emerge above all from the moral field: Simmel always approached the description and definition of social mechanisms and actions in a moralistic rather than a theoretical sense (Roberts 1996: 15). Social philosophy, which was taking shape with Simmel's contributions, is, on the other hand, a collection of elements coming from psychology, anthropology, ethnology, moral philosophy, economics, etc. In the work *Soziologie* Simmel argues:

> The intuition that man is, in all his essence and in all his manifestations, determined by the fact of living in reciprocal action with other men, must certainly lead to a new form of consideration in all the so-called sciences of the spirit (Simmel 1908, It. trans. 1989: 6).

The question of the treatment of individuality had already been effectively addressed by Simmel in the 1892 volume *Die Probleme der Geschichtsphilosophie*, in which the problematization of the historical object itself is interwoven with the theme of individuality. Simmel identifies the presence of unconscious forces operating in the choices and historical dynamics, to be clearly kept in mind in the evaluation of the event under consideration.

> The hypothetical character of every external historical event, which derives from its being possible only by psychological means, is not revealed only by the specific content of the processes of consciousness and by the identical degree of verisimilitude that can be reached by completely opposite suppositions. Indeed, the ambiguity begins already when we ask ourselves where the visible happening is founded by the consciousness and where instead it springs from unconscious forces [...] The unconscious motivation is in fact only the expression of our ignorance of the real operating one; it means only the lack of a conscious motivation (Simmel 1892, It. trans. 1982: 18-19).

The relationship with the historical object is based on the law of understanding, of *Verstehen*. Simmel writes:

> No science is a precise copy of the intact reality of its object, but rather a projection of it on a new level, that is, a reproduction that, while maintaining a continuous relationship with reality, borrows its means and its categories from the specific needs and conditions of the scientific problem, which, compared to the immediacy of the object, appears on the one hand more analytical and on the other more synthetic (Ibid: 117).

Sociology, according to Simmel, must be something other than a pure method applied to the other sciences (reduction of the individual to the social) or a pure new word applied to the totality of the sciences of the spirit.

The concept of sociology in the strict sense of the word must be based on the distinction between the subject matter and the form of a society. Sociology, therefore, as a science of modern society, was born and developed as that form of thought which was supposed to help society to reflect on itself. In this sense, from the very beginning, it has had the vocation to be not only science, but also "consciousness" (moral) of modernity (Roberts 1996: 16).

The object of sociology is the forms of mutual influential relationships that exist between men. This object, society, emerges only and to the extent that several individuals enter into reciprocal action. Society is interaction and the notion of reciprocity is flanked by another fundamental concept of Simmel's sociology: that of "sociation" (*Vergesellschaftung*), which is the process through which a form of mutual action is consolidated over time (Jedlowski 2009: 103-104). Therefore, for Simnel sociology was a formal science, that is, it deals with describing the forms that reciprocal relationships take on in different situations and at different times, solidifying in large institutions, and investigating the way in which relationships between individuals are constituted as social phenomena (Ibidem).

At the beginning of the 20th century, in describing his contribution to the foundation of sociology as a science, Simmel stated:

> In so far as it relies on the consideration that man must be understood as a social being and that society is the bearer of every historical event, [sociology] does not contain any object that is not already dealt with in one of the existing sciences, but is a scientific method that does not constitute a science in itself (Simmel 1908: 7).

Simmel is highlighting the fact that the object that sociology deals with already exists and is studied by other sciences (economics, history, psychology, politics and so on); according to him, this discipline introduces a different perspective, a new way of looking at the same object, so in its relationship with the existing sciences it turns out to be a new method – an auxiliary instrument of research – to approach phenomena in a different way. Sociology must also be a particular point of view, a human reality, which can be analyzed from several points of view. In other words, it must find its cognitive categories. The legacy of Kant in the author's thought is clearly evident; Simmel asks "how is society possible?" and seeks the answer in identifying the cognitive categories of sociology (Izzo 1991: 168).

The general premise that makes sociology a discipline that provides a new approach to events, facts and structures covered by other sciences is that:

> Man, in all his essence and in all his manifestations, is determined by the fact of living in reciprocal action with other men [...]. It is no longer possible to explain historical facts, in the broad sense of the word, that is, the contents of culture, the types of economy, the norms of morality starting from the individual man, his intellect and his interests and, where this fails, to resort immediately to metaphysical or magical causes (Simmel 1908: 6-7).

The analysis of the historical-social object does not aim at inserting the fact within an ideally rational process or one that responds to a teleological principle: the approach to history is always "individual", both in the sense that the historical event is determined by concrete subjects (individuals, social groups, institutions, common people, etc.), and in the sense of its uniqueness and unrepeatability.

> The historically significant happening can be described as the history of the individuals of action – but also as the history of the contents of the happening, i.e. of impersonal, situational or ideal objectivity. The history of art can be written as the history of artists, but also as the history of stylistic forms, the means of artistic expression [...] what is called universal history [...] can be expressed as the history of individual subjects (personal and social) as well as the history of ideas (Simmel 1892, It. trans. 1982: 51).

There is always a double interpretation to explain a social-historical phenomenon, precisely because its analysis should always move, according to Simmel, both in a psychological sense (understanding the individual side) and in a logical sense (that is, taking into account the cognitive structures through which the event is catalogued), thus transforming the psychological data into an "objectively valid" content or meaning. For Simmel, the specific feature of the historian lies precisely in giving the individuality of the object the character of universality: the historian in this is a bit like an artist, who transforms the causality of lived experience (*Erlebnis*) into a universally valid form of happening (Roberts 1996: 15).

An essential element in the configuration of social relationships is conflict; indeed, Simnel highlights the connection between sociology and history through conflict, giving it a positive connotation compared to Marx. Simmel states: «it is not only a means for the preservation of the overall relationship, but it is one of the concrete functions in which it actually exists» (Simmel 1908, It. trans. 1989: 217).

Simmel means that the opposition tests the vitality and strength of the mutual action that is the engine of the association. It is not a question of direct damage to the other, but of the possibility (*a priori*) that society has in the inter-individual configuration, that is, it is necessary for the order of society to be governed by a conflictual and oppositional logic that, in reality, unites by dividing and separating by uniting. "Contrast relationships" can generate or produce a social formation: this is what Simmel means (Ibidem).

The Berlin sociologist highlights three different types of conflict (understood as a form of social relationship): «there are three kinds of conflict relationships existing in society, namely, antagonism, legal conflict, and conflict of interests» (Judge 2012: 126).

Conflict as a "social form" manifests itself as a vital primary social relationship insofar as dynamics – aimed at integration – and universality (group identity) and dynamics – aimed at the need to highlight the differences and distinctions of a complex social system – are activated simultaneously. Moreover, the more complex the social system of reference, the more these dynamics will increase in proportion. According to Simmel, conflict defines the boundaries between the groups within the same social system: in this sense, the internal awareness of the group increases and establishes its identity within the system. The mutual contrast that is generated between the groups actually acts as an element of balance between them, preserving the social system in its entirety (Coser 1967: 41).

However, it was Simmel himself who gave important indications on this subject, when in the pages of his masterpiece *Soziologie*, he wrote:

> that the struggle has a sociological meaning, in that it causes or modifies communities of interests, unifications, organizations, is never contested in principle. It may instead seem paradoxical to the common way of looking at the question of whether the struggle itself, regardless of the manifestations that follow or accompany it, is a form of association (Simmel 1908, It. trans. 1989: 213).

Simmel also understood the struggle as a fundamental expression of human reciprocal action: society needs a quantitative relationship between «harmony and disharmony, association and competition, between favor and disadvantage, to reach a certain configuration» (Ibid: 214). Here is how he describes the struggle and the conflict referring to examples of a historical nature:

> [...] The subjectivity of the final purpose is wonderfully intertwined with the objectivity of the final purpose; it is a supra-individual unity of an objective or social nature that includes the parts and their struggle; you fight with your opponent without turning against him, so to speak without touching him. Thus the subjective antagonistic spring leads us to realize objective values, and the victory in the struggle is not quite the outcome of a struggle, but is the outcome of the realization of the values that go beyond the struggle itself (Ibid: 244-245).

> The conflict that is even in the school in which the ego is formed and the more unitary the ego is to be formed, the more conflict-rich it will be. The more deeply every impression sinks into the inner life, the more conflicts will arise, but the more man will feel unity. Conflict is precisely the form in which the ego is formed and opposed to the world. The soul is where struggle and peace take place (Simmel 1913, It. trans. 1968: 87).

"Life as life requires form", wrote Simmel in *Lebensanschauung*; life needs a form and the relationship that characterizes this need is to be taken as the main object of sociology, while the relationship with the historical object indicated by him is based on the law of understanding, promoting an approach to history that is always of an "individual" character, taking shape in an intersubjective dimension. His formal sociology – his geometry of social space – provided a preliminary map that allowed subsequent researchers to indicate and even predict the movements of social actors as they attempt to transcend those group relationships in which they are located (Coser 1983: 310).

Contemporary sociology possesses technological means and a conceptual apparatus considerably superior to those which were available at the time of Simmel. Nevertheless, it is impossible to say that this author is outdated, or whether he is directly read or filtered through other authors: Simmel continues to stimulate the our sociological imagination in a way that is as relevant as Marx or Weber. The past must never be forgotten, because "the classics are the reserve of the future" (Pontiggia 1998: 58).

5.3 *Max Weber: The relationship between historical and sociological knowledge*

Max Weber focused his attention on the study of social realities starting from the sense that people attribute to their experiences, and he believed that sociology is based on understanding the meaning of human activity.

When he died suddenly in June 1920, Weber left unfinished a huge amount of manuscripts, destined for one of his most important works *Economy and Society*. In this collection of articles and essays Weber summarizes the results of his methodological and theoretical reflections and his sociological, legal, and economic studies. Assuming from the outset an attitude of radical criticism towards theories that consider society as an autonomous system with regard to the action of individuals, Weber observes that in other sciences such as biology, the different concepts of organism and function may be sufficient, while in sociology – whose purpose is to interpret human action – such concepts, albeit important for practical illustration, risk leading to false realism (Weber 1922, It. trans. 1961, vol. I: 13).

Weber considered that the specificity of the contribution of sociology compared to other natural sciences lies in the fact that apart from the simple determination of functional connections and rules, or laws, it is aimed at understanding the attitude of individuals participating in social formations (Ibid.: 14). Although Weber was aware of the advantage of this interpretative approach, he acknowledged the hypothetical and fragmentary nature of its results. Indeed, the limit of sociological science is the impossibility of the encompassing interpretation resulting in a global theoretical system founded on rigid causal connections.

While in the case of animal societies an analysis of the functions related to the logic of conservation and reproduction of life may be appropriate, in the case of human societies, according to Weber we not only face processes of differentiation connected to instinctive or mechanical structures, but actions that are the product of a conscious rationality. This particular dimension can only be highlighted by a form of sociology that proposes to understand the sense in which behaviour is determined (Crespi 2002: 50).

In other words, in *Economy and Society* Weber provides this definition: «sociology must designate a science which proposes to understand by virtue of an interpretative procedure of social action, and therefore to explain it, causally in its course and its effects» (Weber 1922, It. trans. 1961, vol. I: 12).

Therefore, sociology is first and foremost a science that interprets and understands social action. However, understanding is not a synonym for explaining; in fact, the causal explanation comes after the application of the interpretative procedure.

In order to better understand what has been said so far, it is useful to emphasize that for Weber to understand an action means understanding its meaning, that is interpreting the meaning that that action has in the eyes of the person who carries it out. Social action, the object of sociology, is an action endowed with meaning. An act can be considered such, therefore, if and in so far as there is a meaning connected to it, that is, the meaning that the person who carries out the action attributes to the action itself.

The possibility of understanding distinguishes the social and human sciences from the natural sciences[1]. Weber believed that while in the natural sciences phenomena are not agitated by subjects that give them meaning, in the human sciences, the scientist faces phenomena that are agitated by subjects who attribute a meaning to them.

For example, if a stone falls, one can describe the motion and explain it by referring it to hypothetical general laws that govern it, but one should not question the meaning of falling for the stone itself, since the stone itself has no conscience. But if a man throws a stone, what the scientist can and must do first is to understand the meaning of that gesture. This is possible because the man who throws the stone is a subject, who acts with motives and ends. He can throw the stone for fun, because he is having a throwing contest or because he wants to hurt someone. The meaning of the action is different in each of these cases and if one cannot grasp these differences, the gesture remains obscure.

Therefore, Weber considers all social sciences to be inclusive sciences; that is, sciences whose object is action as behaviour endowed with meaning. However, there are differences between the different scientific disciplines. History, in particular, dealing with the uniqueness of events, aims to understand facts that have occurred only once, and is marginally interested in the regularity with which the phenomena are manifested. On the contrary, sociology, as a science oriented to generality, aims to study the social actions of men regarding what is typical about them, i.e., the characteristics that recur in several cases. Because of this orientation, sociology must abstract certain common characteristics from infinite singular actions and produce types of phenomena.

---

[1] This point is fundamental, because it represents a break with the approach adopted by the earliest Enlightenment thinkers up to Durkheim. In fact, from Montesquieu to Comte and up to Durkheim, the scientific model *par excellence* is that of the natural sciences, and the human sciences must progressively adapt to this model (Jedlowski 2009:124).

As mentioned above, sociology does not only aim to understand, but also to explain the action causally, and this for Weber means characterizing in a modest way some factors that have conditioned the emergence of determined phenomena that they aim to explain and that are always present when they manifest themselves. Therefore, identifying a cause of a given phenomenon does not mean finding the causal factor that determines it in a necessary way, but only identifying the relevant and appropriate conditions to explain it.

It is interesting to note that natural sciences proceed in principle by trying to give causal explanations of the phenomena they study. Once the sociologist has understood the meaning of the phenomena he/she observes, his or her next steps are therefore similar to those of the natural scientist, in that he/she also seeks causes that explain the onset of those phenomena.

Weber considers that for human phenomena there is no perfectly exhaustive causal explanation, as there are so many factors that combine to produce a phenomenon that it cannot be established that a given phenomenon is caused by a given factor.

For example, if you want to trace the causes of phenomenon B: if every time that B appears, the phenomenon A has previously appeared, it is reasonable to suppose that B is caused by A. But if A appears sometimes even without the appearance of B, then it can be concluded that A is not an adequate cause of B, but that it is probably a factor that facilitates its occurrence, but it alone is not sufficient to explain it. But assuming that A always appears only when B manifests itself, then in this case, it will be an appropriate cause of B. The hypothesis remains opened since we have not noticed another phenomenon called C, which perhaps has an even more important role than A to causally explain B. This distinguishes social science research. According to Weber, social reality is nothing more than an infinitely thick and extensive mesh fabric, so you cannot be absolutely sure you have found exhaustive causal explanations to any phenomenon (Jedlowski 2009: 123-125).

> A decision even of the smallest section of reality can never be thought [as exhaustive] [...] The number and type of causes that have determined any individual event is, in fact, always infinite, and there is not a characteristic inherent in things to isolate a part of them, which is taken into consideration alone [...] Only certain aspects of particular phenomena, always infinitely multiple, that is, those to which we attribute a cultural meaning [...] are therefore worthy of being known, and they alone are the subject of the causal explanation (Weber 1904, It. trans. 1958: 92-93).

The identification of causal connections always depends to a certain extent on the choices of the scientist who, thanks to his experience, his own ideas and his own orientations will be more able to see certain causal connections than others.

This is why Weber prefers to talk not about causes but about conditions or set of factors (Jedlowski 2009: 123-125).

In order to better understand the contents presented, in the early part of *Economy and Society* Weber further defines the specific scientific function of sociology. Indeed, we can read:

> Sociology elaborates – and this has been repeatedly taken as evident – concepts of types and seeks general rules of becoming in antithesis to history, which aims at causal analysis and imputation of actions, of individuals such as those of cultural importance. The conceptual elaboration of sociology draws its material – in the form of models – essentially, though not exclusively, from the realities of action which are also relevant from the point of view of history. It forms its concepts and investigates in search of rules – Weber specifies – especially also according to the perspective of the usefulness that they can, for this reason, reveal in view of the historical-causal imputation of the phenomena of cultural importance (Weber 1922, It. trans. 1961, vol I: 17).

This means that the fundamental question is precisely the formulation of general rules through specific conceptual elaborations of the reality of social action (Carbonaro 1986: 14). Therefore, it is not so much the action of the individual, but the action of the individual type of a multiplicity of holders of the action, which is inserted as a social action within the historical segment of culture.

In analyzing the relationship between history and sociology, Weber believes that sociology must provide both history and the historian with the useful tools to carry out its scientific function, that is, to identify specific causal links that connect different phenomena relevant on the cultural and decisive level, the historical one, and the individual one (De Nardis 1999: 26-27). So what for Weber is the historical-causal attribution of phenomena of cultural importance would seem to rest on the one hand on a conceptual theoretical framework as a reference parameter, and on specific knowledge in the form of general laws governing social action, which sociology offers history on the other hand (Cavalli 1981: 43). The former is oriented to the study of behavior uniformity and their causal connections, and the latter is aimed at the explanation of individual phenomena. This does not mean, however, that sociology must turn its gaze to that search for general laws of society and history that had marked its first affirmation as an autonomous discipline, in France with Comte, and in

England with Spencer. In fact, sociological knowledge has its own autonomy and specificity, but it is a continuous interaction with historical knowledge and must extract from historical-empirical material generalizations limited in space and time, which in turn serve to guide the work of historical and empirical research (Weber 1993: 13-14).

In Weber's perspective, sociology and history are interconnected by virtue of some methodological principles that these two distinct disciplinary fields share: selection in relation to the values, the construction of historical individualities, and the explanation as an adequate causal relationship – that is, probabilistic and partial. Although the aims of the historical method and sociological analysis are different, Weber believes that history and sociology must be considered mutually integral; whereas historical understanding requires the use of propositions of a general nature which historical comparisons alone cannot demonstrate, sociological explanation aspires to establish regularity which its conceptual categories are not sufficient to prove.

Thanks to his ability not to separate the study of social structures from the observation of the multiple forms of individual action, Weber defines a theoretical perspective in which the individual and social levels of sociological analysis are closely integrated. Thus on the one hand the historical-comparative examination of the great social systems in the form of complex collective orientations, and on the other hand, the conceptual typing of the relationships that exist between these systems and the intentions that guide the meaningful action of individuals, constitute for Weber the most appropriate scheme for the explanation and understanding of the social action.

The importance attached by Weber to the concepts of action and social relationship within his comprehensive sociology is a key element in explaining legal, economic and religious systems that the author himself develops in the context of his historical-comparative sociology, understanding them as the product of the action with the sense of subjectively intentional individuals (Fasano 2012: 103).

Therefore, both sociology and history apply the reference to the value, both aiming to build ideal types and they have the material in common; however, while in history the ideal type grasps what is specific, what makes a phenomenon or an event important in relation to certain cultural values that make it meaningful for anyone who accepts it, in sociology, the ideal type expresses laws that regulate the essential aspects of the behaviour of individuals. Another element of difference between sociology and history is that while the former deals with the essentials in established, stable life relationships, at least for a certain period, the latter deals with what is essential in phenomena and events that are not common and indeed change a tradition.

History studies the emergence of novelty. Within social action there can be observed uniformity of fact, that is, processes of action which are repeated in the same individuals acting or which extend to numerous individuals with a typically homogeneous intention. Sociology deals with these types of proceeding of action, in antithesis to history, which tends to the causal imputation of important individual connections, fraught with consequences (Weber 1922, It. trans. 1961, vol. I: 26).

Finally, from a strictly methodological point of view the only difference between history and sociology is that of extension. Indeed, while the former investigates all the planes of action, the latter studies the relationship between means and purposes, leaving aside the reference to cultural values (Catarzi and others 1987: 50-51).

The historical and sociological sciences both aim to explain causally and to interpret in a comprehensive way.

The sociologist not only theorizes the system of beliefs and behaviour of the community, but also wants to establish how things happened, how a certain way of acting is determined by a certain way of believing or how a certain political organization can influence the organization of the economy. The analysis of causal determinations is one of the procedures that guarantee the universal validity of scientific results.

According to Weber, causal research can be oriented in two directions, historical causality and sociological causality. While the former creates unique circumstances that have caused a certain occurrence, the latter assumes the existence of a regular relationship between two phenomena that does not necessarily take the form in which phenomenon A makes B inevitable, but phenomenon A more or less favours phenomenon B. An example of this may be the phrase (true or false as it may be) "a despotic regime favours the intervention of the State in the management of the economy".

The problem of historical causality concerns the determination of the part represented by the different antecedents in the birth of an event and supposes some steps. First of all, it is necessary to construct the historical individuality of the event whose causes are claimed to be found, as this allows us to identify with precision the characteristics of the event; subsequently, all the elements that constitute the historical phenomenon are analysed and finally, if we consider a particular event which happened only once, in order to arrive at a causal determination, after analysing the historical individuality and its antecedents, we should assume that one of the antecedents did not occur or occurred in a different way.

In other words, one should ask oneself "what would have happened if?" Having done this, it is necessary to compare the unreal becoming, which was constructed from the hypothesis of the modification of one of the antecedents, with the real evolution, in order to arrive at the conclusion that the element modified by the thought was one of the causes of the historical individuality found in the principle of research. In this way, the dimension of uncertainty or probability that characterizes events – as well as any man of action lives and conceives them – is given to the events of the past.

The more general propositions the historian has at his disposal that will make it possible to construct unreal evolutions, and to specify the probability of a certain event in relation to one of the preceding ones, the most rigorous the historical causality will be.

In this regard, Weber considers that there is a connection between historical causality and sociological causality, as both express themselves in terms of probability. So Weber believes that the causal relationships of sociology are improbable and partial relationships, in the sense that a fragment of reality makes another fragment of it probable or improbable.

In conclusion, Weber refuses to contemplate that the sciences that have as their object human reality are exclusively – or even only mainly – historical sciences. It is true that the sciences of human reality are more interested in what is unique, or in becoming unique, than the natural sciences; however, they do not neglect the general propositions. The sciences of human reality are not sciences, even when they aim to understand what is particular, except to the extent that they are able to establish general propositions. According to Weber, there is therefore an intimate relationship between the analysis of events and the formulation of general propositions.

History and sociology are two disciplines that should acknowledge each other, because historical understanding requires the use of general propositions and these cannot be proved only by starting from analysis and historical comparisons (Aron 1989: 466-473).

This solidarity between history and sociology can be seen above all in the concept of the ideal type, which is the soul of Weber's epistemological doctrine:

> The ideal type represents a conceptual framework, which is not historical reality, nor the "real" reality [...]: it has the meaning of a pure ideal concept-limit, to which reality must be commensurate and compared, in order to illustrate certain significant elements of its empirical content (Weber 1904, It. trans. 1958: 12).

In other words, the ideal type recalls this fluid conception of the relationship between historical knowledge and sociological knowledge for several reasons: the ideal type is linked to the concept of understanding, from this relationship in which each ideal type is an organization of intelligible relationships pertaining both to a historical whole and to a succession of events; it is linked to what is distinctive of the society; the ideal type is also linked to the analytical and partial concept of causality, allowing us to perceive historical individuals or historical sets; and finally the ideal type is the main tool to understand a particular type of behavior or historical phenomenon.

In conclusion, historical or sociological causality occupies a fundamental role in the thought of Weber, who, being immersed in the intellectual climate of German Historicism, on the one hand shared the importance of subjectivity (hence the study of social realities starting from the sense that men attribute to their experiences), and on the other hand rejected the idea that social science can stop at the description of the details, reducing itself to a collection of included events. For Weber, history is a continuous evolution and for this reason there is no certain rule that goes beyond the action of man, while sociology must understand human experiences in their distinctiveness, yet at the same time derive "general models" from that study (Aron 1989: 466-476; Cesareo 2006: 27).

Bibliographical references

Abd el-Malek, A., *La dialettica sociale*, trans. by G. Barletta (Bari: Dedalo Libri, 1974).
Aron, R., *Le tappe del pensiero sociologico. Montesquieu, Comte, Marx, Tocqueville, Durkheim, Pareto, Weber*, trans. by A. Devizzi (Milano: Mondadori, 1989).
Bourdieu, P., and L. Wacquant, *Réponses. Pour une Anthropologie Réfléxive* (Paris: Seuil, 1992); It. trans. Bourdieu, P., *Risposte per un'antropologia riflessiva* (Torino: Bollati Boringhieri, 1992).
Burke, P., *Sociologia e storia* (Bologna: Il Mulino, 1982).
Busino, G., *Sociologia e storia: elementi per un dibattito* (Napoli: Guida, 1975).
Carbonaro, A., *La legittimazione del potere: problematica del rapporto tra stato, istituzioni e società* (Milano: Franco Angeli, 1986).
Catarzi, M., Centi, B., Franco, V., Meschiari, A., Orsucci, A., and M. Turchetto, *Disincanto e ragione. Filosofia valori e metodo in Max Weber* (Bari: Edizione Dedalo, 1987).
Cavalli, L., 'Osservazioni sulla «natural science sociology»', *Studi di Sociologia*, 3 (1963), 213-227.
— *Il capo carismatico: per una sociologia weberiana della leadership* (Bologna: Il Mulino, 1981).

Cerroni, U., *Materialismo storico e scienza* (Lecce: Milella, 1976).
Cesareo, V., and I. Vaccarini, *La libertà responsabile. Soggettività e mutamento sociale* (Milano: Vita e pensiero, 2006).
Collins, R., and M. Makowsky, *Storia delle teorie sociologiche* (Bologna: Zanichelli, 1980).
Coser, L. A., *Le funzioni del conflitto sociale* (Milano: Feltrinelli, 1967).
— *I maestri del pensiero sociologico* (Bologna: Il Mulino, 1983).
— *I maestri del pensiero sociologico* (Bologna: Il Mulino, 1997).
Costantino, S., and A. Zanca, *Sociologi: tra moderno e post-moderno* (Milano: Franco Angeli, 2015).
Crespi, F., *Il pensiero sociologico* (Bologna: Il Mulino, 2002).
D'Anna, V., *Max Scheler. Fenomenologia e spirito del capitalismo* (Roma: Editori Riuniti, 2006).
De Nardis, P., *Sociologia del limite* (Roma: Meltemi Editore, 1999).
Del Vecchio, G., *Critica del materialismo storico* (Roma: Angelo Signorelli, 1964).
Fasano, L., *Teoria della scelta razionale e individualismo metodologico. Un riesame critico* (Milano: Egea, 2012).
Ghisleni, M., *Teoria sociale e modernità. Saggio sulla storia della sociologia* (Roma: Carocci, 1998).
Giddens, A., *Capitalismo e teoria sociale. Marx, Durkheim e Weber*, ed. by A. Martinelli, trans. by C. Cantini and M. Pogatschnig (Milano: Il Saggiatore, 2009).
Izzo, A., *Storia del pensiero sociologico* (Bologna: Il Mulino, 1991).
Jedlowski, P., *Il mondo in questione. Introduzione alla storia del pensiero sociologico* (Roma: Carocci, 2009).
Judge, P., *Foundation of Classical Sociological Theory. Functionalism, Conflict and Action* (London: Pearson, Dorling Kindersley, 2012).
Magnano San Lio, G., *Forme del sapere e struttura della vita. Per una storia del concetto di Weltanschauung. Tra Kant a Dilthey* (Catanzaro: Rubbettino, 2005).
Marx, K., *Oekonomisc-Philosophische Manuskripte aus dem Jahre* (Frankfurt: Suhrkamp, 1844a); Eng. trans. 'Economic and Philosophical Manuscripts', in K. Marx, *Karl Marx: Selected Writings*, ed. by D. McLellan (Oxford: Oxford University Press, 2000), pp. 83-121.
— *Oekonomisc-Philosophische Manuskripte aus dem Jahre* (Frankfurt: Suhrkamp, 1844a); Eng. trans. *Economic and Philosophic Manuscripts of 1844*, trans. by M. Milligan (ed.) (New York, Mineola: Dover Publication Inc., 2012).

— *Zur Kritik der Hegelschen Rechtsphilosophie* (Paris: Deutsch–Französische Jahrbücher, 1844b); Eng. trans. 'Introduction to a contribution to the critique of Hegel's "Philosophy of right"' in K. Marx *Critique of Hegel's "Philosophy of right"* (New York: Cambridge University Press, 1970), pp. 129-142.

— *Grundrisse der Kritik der politischen Ökonomie* (Rohentwurf, 1857-58); It. trans. 'Lineamenti fondamentali della critica dell'economia politica (1857-58)', in K. Marx, *L'alienazione*, ed. by M. Musto (Roma: Donzelli, 2010), pp. 43-82.

— *Zur Kritik der politischen Ökonomie* (Berlin: Duncker, 1859); Eng. trans. *A Contribution to the Critique of Political Economy*, trans. by N. I. Stone (Chicago: Charles H. Kerr and Co., 1904).

Marx, K., and F. Engels, *Die deutsche Ideologie* (Berlin: Dietz, 1845-46); Eng. trans. *The German Ideology (Part one)*, ed. by C. J. Arthur (New York: International Publisher, 1970).

— *Die deutsche Ideologie* (Berlin: Dietz, 1845-46); Eng. trans. 'The German Ideology', in K. Marx and F. Engels, *Collected Works*, trans. by R. Dixon and others (New York: International Publishers, 1976), vol. V.

— *Die deutsche Ideologie* (Berlin: Dietz, 1845-46); Eng. trans. 'The German Ideology', in R. C. Tucker (ed.), *The Marx-Engels Reader*, 2nd edn (New York: W. W. Norton, 1978), pp. 146-200.

— *Manifest der Kommunistischen Partei* (London: Burghard, 1848); Eng. trans. *The Communist Manifesto* (London: Penguin Books, Harmondsworth, 1985).

— *Sul materialismo storico* (Roma: Edizioni Rinascita, 1949).

Meschiari, A., 'Moritz Lazarus e Georg Simmel', *Giornale critico della filosofia italiana*, I (January-April 1996), 52-82.

Mills, C. W., 'L'uso della storia', in P. Crespi (ed.), *Sociologia e Storia* (Milano: Celuc, 1974), pp. 191-216.

Pontiggia, G., *I contemporanei del futuro. Viaggio nei classici* (Milano: Mondadori, 1998).

Roberts, D., 'Georg Simmel's Philosophy of Money. Reflexions on the relation between Philosophy and History', *Thesis Eleven*, 44 MIT (1996), 12-27.

Schnabel, E., 'Georg Simmel', in von D. Käsler, *Klassiker des soziologisches Denkens* (München: Beck, 1976), vol. I , pp. 267-311.

Simmel, G., 'Über sociale Differenzierung' (1890), in *Aufsätze 1887 bis 1890. Über sociale Differenzierung (1890). Die Probleme der*

Geschichtsphilosophie (1892); It. trans. *Sulla differenziazione sociale* (Bari: Laterza, 1982).
— 'Die Probleme der Geschichtsphilosophie' (1892), in *Aufsätze 1887 bis 1890. Über sociale Differenzierung (1890). Die Probleme der Geschichtsphilosophie (1892)* (Frankfurt: Suhrkamp, 1989); It. trans. *I problemi della filosofia della storia* (Casale Monferrato: Marietti, 1982).
— *Soziologie: Untersuchungen über die Formen der Vergesellschaftung* (Berlin: Dunker & Humblot, 1908).
— *Sociolozie, Untersuchungen über die Formen der Vergesellschaftung* (Berlin: Duncker & Humblot, 1908); It. trans. *Sociologia*, ed. by A. Cavalli (Milano: Comunità, 1989).
— *L'etica e i problemi della cultura moderna*, ed. and trans. by G. Calabrò (Napoli: Guida Editori, 1913, It. trans. 1968).
— *Die Grundfragen der Soziologie*, in *Der Krieg und die geistige Entscheidungen. Grundfragen der Soziologie. Vom Wesen des historischen Verstehens. Der Konflikt der modernen Kultur*, GSG 16 (Frankfurt: Suhrkamp), 1917, pp. 84-85; It. trans. *Forme e giochi di società. Problemi fondamentali della sociologia* (Feltrinelli: Milano, 1917).
— *Arte e Civiltà*, trans. by D. Formaggio and F. Perucchi (eds) (Milano: Isedi, 1918, It. trans. 1976).
— 'Vom Wesen des historischen Verstehen (1918)', in *Der Krieg und die geistige Entscheidungen. Grundfragen der Soziologie. Vom Wesen des historischen Verstehens. Der Konflikt der modernen Kultur*, GSG 16 (Suhrkamp, Frankfurt: Suhrkamp, 1977, repr. 2000), pp. 151-179; It. trans. *L'essenza del comprendere storico*, in *Lo storicismo tedesco* ed. by P. Rossi, (Torino: UTET), pp. 511-537.
Weber, M., 'Die "Objektivität" sozialwissenschaftlicher und sozialpolitischer Erkenntnis', Archiv für Sozialwissenschaft und Sozialpolitik, 19 (1) (1904), 22-87; It. trans. 'L'oggettività conoscitiva della scienza sociale e della politica sociale', in *Il metodo delle scienze storico-sociali*, ed. by P. Rossi (Torino: Einaudi,1958).
— *Wirtschaft und Gesellschaft* (Tübingen: Mohr, 1922); It. trans. *Economia e Società*, 2 vols (Milano: Edizioni Comunità, 1961).
— *Storia Economica. Linee di una storia universale dell'economia e della società. Introduzione di Carlo Trigilia* (Roma: Donzelli Editore, 1993).

SOCIAL AMBIVALENCES AND CONFLICTS
IN EMPIRICAL SOCIO-HISTORICAL RESEARCH

Ambivalences and Conflicts
in Socio-Historical Process

Tova Benski

6.
"CHANGE BEGINS WHERE SILENCE IS BROKEN".
DYNAMICS OF CONTINUITY AND CHANGE IN THE WOMEN'S PEACE MOVEMENT IN ISRAEL 1983-2010

In the concluding chapter of her book *The Israeli Peace Movement: A Shattered Dream* (2009), Tamar Herman stated that throughout the long years of peace activism in Israel, the peace movement had been successful in changing government policy in only two peace initiatives. One of them was the withdrawal of the IDF from South Lebanon in mid-2000, following the activities of *Four Mothers*, a women-only peace movement. *Four Mothers* (symbolically pointing to the four biblical mothers of the nation) that emerged in Israel in 1997, following the crash of two helicopters carrying Israeli soldiers to Lebanon in which 73 young Israeli combat soldiers were killed.

Despite the widely acknowledged credit given to *Four Mothers*, the wider arena of women's peace movements in Israel has received no acknowledgement and almost no media coverage during their long years of persistent activism. I have always seen this as a remarkable example of the marginalization and transparency of these mobilizations. As I saw these mobilizations change from decade to decade, I was intrigued by the fact that they were non-existent in terms of social and political visibility. I also realized that most of the Israeli public were unaware of the women's peace camp in Israel. This made me ponder the question of historical justice regarding women peace activists. I felt that, for the sake of both present and future generations of women peace activists, I need to give this part of the Israeli peace camp visibility and voice that has been denied them for many years and to correct this historic (or *herstoric*) wrong.

This presentation is based on an almost lifelong study of women's peace mobilizations in Israel. I entered the field in 1985 and am still on and off in the field. The data was collected through a triangulation of methods including: participant observations at meetings; vigils, demonstrations, marches, committees, conferences, and other activities organized by different movements/organizations; emails and reports of activities; group and individual in-depth, open-ended interviews at different points along the time span of their activism; self-administered questionnaires; and many other sources such as articles, publications, memoirs and other materials appearing on the movements' websites, Facebook, Twitter and other media. During my 35 years of involvement in activism and research in the field, I have collected enormous amounts of data and notes, and have had the unique opportunity to watch, experience and detect continuities and changes along both the horizontal and the time axes. Some of these I will share with you in this presentation. However, before we start, some notes about the context are needed.

6.1 *Silencing women's voices in modern Israeli society*

From a feminist perspective, Israeli modernity has endowed women with numerous rights but, at the same time, forced them into a position of continuing social and political marginality[1]. Feminist scholarship in Israel identifies three core factors that are held responsible for the exclusion of women from the political centers of power and from the security discourse, which remains the dominant public discourse: a. the private/public binary and the gendering of this dualism (Azmon 1995; Herzog 1994, 1999, 2003; Naveh 1999;); b. the absence of a clear separation between the state, religion, and patriarchal tradition originating in ancient Jewish tradition and religion (Azmon 1993; Elior 2001); and c. the continuing Israeli-Palestinian armed conflict[2]. These three factors, fundamental to the development of Israeli modernity, have together effectively positioned women at the margins of society and contributed to their subjugation to patriarchic control practices (Elior 2001; Radai, Shalev and Liban-Kobi 1995).

For the current discussion, the most relevant factor is the ongoing Israeli-Palestinian conflict. In very short and even simplistic terms, years before the establishment of the state of Israel in 1948, armed conflict had been raging between Jews and Arabs over the land and the right to establish a Jewish state. This conflict erupted into wars, operations, attacks on civilian targets by both sides of the conflict, invasions, and occupation of Palestinian territories by the IDF (Israeli Defense Forces) (1967). Despite attempts to reach a resolution of this conflict, it is still raging, and Israel has become an embattled society in a constant state of military readiness. The centrality of the conflict in Israeli reality finds expression in the collective identity. Due to the centrality of the assumption of existential threat posed to the

---

1 See Azmon 2001; Elior 2001; Fogiel-Bijaoui 1997; Herzog 1994, 2003; Radai, Shalev and Liban-Kobi 1995.
2 See Azmon 1993; Berkovitch 1999; Herzog 2003; Smooha 2003.

state of Israel by the surrounding Arab countries, military considerations and issues defined as "national security" take the highest priority on the national agenda. These considerations became part of the organizing principles of the collective. All other issues were subjected to secondary position. These characteristics are at the heart of various definitions of Militarism (Kimmerling 1993; Ben Eliezer 2003).

The centrality of the conflict has affected the position of both men and women in Israeli society. The construction of masculinity and femininity in societies mobilizing for war is intertwined with the gendered duties of fighting versus nursing/nurture (Helman 1999). According to Berkovitch (1999), within Israeli political culture, a series of criteria evolved that measured good citizenship according to the individual's contribution to national security. Hence, military service – especially in combat units – defined good male citizenship, whereas motherhood defined "good women" citizenship. War making and nation building have been associated with the reproduction and sharpening of gender and other divisions (Melman 1998; Reardon 2005). Israel's political culture thus contributed to the gendering of the public/private division, a process that culminated in gender inequality, the exclusion of women from the political-security discourse, and the marginalization of issues that were important for women (Herzog 2003).

History has shown that as early as the beginning of the 20th century, women in Israel resisted their exclusion and marginalization and demanded integration within the community's political and economic institutions (Bernstein 1987; Fogiel-Bijaoui 1992; Safran 2001). At the same time, many women (at times, reluctantly) accepted the centrality of the security threat and the developing hierarchy of issues. Women's Peace movements, that have evolved in Israel since the 1980s, broke this consent and the silence associated with it and mobilized to promote both the status of women, their voice in security issues, and to express their disagreement with Israeli government policies in the territories occupied during the 1967 war.

6.2 *Women's peace movements in Israel 1983-2020*

6.2.1 *The Beginning*

Women have long been active in many mixed-gender peace movements like *Peace Now, the 21st Year, Stop the occupation*, etc. The break with these movements and mobilizing into women-only peace movements represents a relatively new phenomenon in Israeli civil society. This trend began during the 1980s with the emergence of various women-only peace movements that have been active on and off since then. Some of these movements disbanded and other new movements appeared on the scene, while some have persisted for more than three decades and are still active, even though their numbers and attendance have dwindled.

As is well known, movements ebb and flow and Women's Peace Movements are no exception. They have evolved through a dynamic of periods of increased mobilization (waves of protest). The emergence and dynamics of women-only peace movements in Israel is inextricably bound with outbursts of violent periods of the conflict. From the first case of mobilization immediately following the outbreak of the first Lebanon War in June 1982, we can now identify three large periods of increased activism (waves) with periods of abeyance in between that have also produced some important developments. The first wave emerged in response to the outbreak of the first Palestinian *Intifada* (uprising) in January 1988 and receded in 1994/1995 with the signing of the Oslo Accords. The second period of increased women's peace activism emerged with the onset of the *second Intifada* (October 2000) and the most central organization, the *Coalition of Women for [Just] Peace*, is still active today. The present third wave consists of the movement *Women Wage Peace (WWP)* which emerged in response to the 2014 operation *Protective Edge* in Gaza and the continued evolving second Coalition.

The historical-sociological account of the first two waves of these movements forms the main objective of this presentation. The descriptive account and analysis follow the historical sequence of the appearance of these movements. First, the "early bird" *Parents/Mothers Against Silence* is presented in brief. Then discussion begins of what is generally referred to as the feminization of Peace protests in Israel. Under this heading we proceed with presenting the movements of the first wave and their characteristics, the movements of the interim period and the movements of the second wave. The chief movements in each of these waves are presented through consideration of the following characteristics: participants, organization, and activity; value orientations and identity; and globalization. The two last parts before the Epilogue present a comparative analysis in terms of continuities and changes between these two waves.

6.2.2 *"Early Bird": 1982-1985 – Mothers Against Silence*

The first women's peace mobilizations emerged in Israel during the first Lebanon war that began in 1982, including *Parent/Mothers Against Silence*, that persisted in its activities for over two years. The initiators and members of the movement that was announced on May 1983, were parents, grandparents and siblings of soldiers involved in combat in Lebanon, who had not been active previously. The participants chose to call themselves *Parents Against Silence,* but the press called them *Mothers Against Silence,* and this name stuck and was adopted by the movement without too much enthusiasm.

Most of the members were Ashkenazi (of European origin) women, mothers of soldiers fighting in Lebanon. In terms of socioeconomic status, they were middle class women with high educational levels and active in the labor force, mostly in professional occupations. The participants did not see themselves as "feminists", or as ideologically left-wing; they did not challenge militarism, nor did they touch the issue of the occupation; they were not adversarial. They were simply critical of the Government's decision to invade Lebanon and defined it as an "unnecessary" war. They had only one goal: the retreat of the IDF from Lebanon and the return of their sons. They claimed that the IDF had been established with the goal of defending Israel in cases of emergency and eminent existential threats on the state and its citizens. A war that was not defensive was framed as a breach of a basic consensual principle of Israeli democracy whereby the only justification for sacrificing life was for defense purposes. For the first time in Israeli history, parents blamed the government for their sons falling in battle and did not accept it as a necessary sacrifice for the existence of the state. The movement turned Republican motherhood into an oppositional symbol and mobilized it to challenge what they called "unnecessary" war (Zukerman-Bareli and Benski 1989; Gilat 1987, 1991; Azmon 1997). In 1985, when it seemed that the army had started withdrawing some of its forces from Lebanon, the women had the impression that they had achieved their goal and so they disbanded.

For our purposes in this presentation, the importance of this "early bird" mobilization lies in two main achievements: 1. legitimizing the right of mothers to protest war; they managed to insert the maternal voice into the discourse over security issues in Israel and paved the way to be followed by *Four Mothers* (1997), and 2. breaking the silence concerning 'unnecessary' wars and calling on parents to break their silence; breaking the silence is a characteristic that has since been common to all the women's peace movements that followed.

6.2.3 *The feminization of peace protests in Israel*

Three years after *Mothers Against Silence* dispersed, many women peace activists left the mixed gender peace organizations and organized themselves into women-only peace initiatives and mobilizations. There are various explanations for this feminization of peace activism (Deutch-Nadir 2005). The three most common explanations are: 1. the conscious decision taken by the Israeli Feminist movement in the early 1980s to avoid the issue of the occupation; 2. their marginalization in the mixed peace organizations[3]; and 3. the special nature of the Palestinian Intifada[4].

The decision of the Feminist movement to avoid the issue of the occupation and oppression of the Palestinians in the territories that were conquered by Israel following the 1967 war led women who wanted to engage the issue of the Israeli-Palestinian conflict to look for other settings outside the movement. Some joined the three women-only organizations that had been established before the first intifada: *Gesher, Gesher Lashalom,* and *WILPF* (a branch of the Women's International League for Peace and Freedom), and *TANDI – the Movement of Democratic Women in Israel*. Others joined mixed-gender organizations, like *Peace Now, The 21st Year, Stop The Occupation,* etc. where they comprised between 60%-90% of the members but were not part of the decision making apparatus; their voice was not heard and they remained invisible (Sasson-Levi 1995; Chazan 1991).

Despite their marginalization in the mixed-gender movements, as was the case in the US with the emergence of the Feminist movement from the ranks of the New Left (Freeman 1973), the women gained experience in organizing and the routine management of a movement, which gave rise to an infrastructure of experienced women. An internal network of communication emerged among them (remember that this was before electronic media, cellphones, and the Internet), and for some women this was a firsthand experience of a journey into the feminist consciousness of the oppression of women and their marginalization. This contributed to the development, and in some cases, radicalization, of feminist consciousness. At the same time, it established an infrastructure of women with experience who were connected in women's networks, ready to mobilize at the first opportunity.

The Palestinian Intifada in the territories occupied by Israel since the 1967 war burst onto the scene in December 1987 and gave rise to a wave of peace activism of unprecedented magnitude in Israeli society. The Intifada served as a catalyst for the political awakening of Jewish Israeli women (Svirsky 2003). It brought the reality of the occupation to their homes as they saw TV coverage of armed Israeli soldiers ("our sons", as some of them said) chasing women and children. This made it clear that the Intifada

---

3 See Deutch-Nadir 2005; Chazan 1991; Benski 2002; Sasson-Levy 1995.
4 See Deutsch 1994; Pope 1993; Mayer 1994; Benski 1998.

was a civilian uprising and that the oppressors were the Israeli army. For the second time since the war in Lebanon and perhaps under the influence of their predecessor – *Mothers Against Silence* – the assumption that the IDF was acting as a defensive army was shattered. The women started questioning the role that the men in their lives (spouses, children, fathers, uncles, and others) fulfilled during their army and reserve service, and the answers were very painful (Svirsky 2003; Deutch-Nadir 2005).

6.2.3.1 *The first wave of women's peace activism in Israel*

1988-1994/5

The first wave of women's peace activism was composed of dozens of organizations, single issue movements, *ad hoc* and more permanent groupings. Some of these were veteran, well-organized groups like *TANDI*, the *Bridge,* and *WILPF*, but most of them were new movements that had emerged in response to the intifada like *Women in Black, Shani, The Peace Cloth/Quilt, Women for Women Political Prisoners,* and others. These movements were active at least till 1994-1995 and slowed down after the Oslo agreement was signed.

Looking at the most active movements in this wave, we can roughly divide the field of women's peace activism in this period into two main wings: the radical *Coalition of Women and Peace* and the more moderate liberal feminists and peace activists of *The Women's Peace Net* who established the *Jerusalem-link* and *Bat-Shalom* and in 1994 merged with Bat-Shalom. For the first time in Israel, an autonomous women's peace movement emerged and shaped its own political agenda, and according to several accounts was the most creative element in this wave of activism (Svirsky 2003).

• *The Coalition of Women and Peace*

The Coalition of Women and Peace emerged in 1988 with the goal of coordinating and widening the scope and impact of peace activities of women and the new organizations by means of combining the forces of peace seeking women in Israel. Seven peace groups joined the coalition *Women in Black, The Bridge, The Bridge to Peace* and the Israeli branch of *WILPF* – *Women's International League for Peace and Freedom, Shani, The Peace Cloth/Quilt, Women for Women Political Prisoners, and TANDI – movement of Democratic Women in Israel.* Each group retained its autonomy while cooperating and coordinating some of their activities. This was the first coalition of women's peace movements/networks in Israel. It was considered radical in its value orientation, activities, and adversarial attitudes towards the government and its political institutions.

The common objectives of the groups within the coalition were: 1. to increase the involvement of women in politics, in feminist issues, and in issues concerning peace and the military occupation; 2. to establish a framework for dialogue and cooperation between Jewish and Palestinian women citizens of Israel, and between Israeli and Palestinian women from the West Bank and Gaza (today the right term would be Palestinian authority); 3. to advance the mutual interests of women on both sides of the Middle East conflict and towards the empowerment of women; and 4. to exert pressure on the Israeli government to enter peace negotiations and end the occupation. (Presented on June 4, 1989, Haifa).

Participants, organization and activities

It is not possible to give a realistic estimate of the size of the participants of each group due to the great extent of overlapping participation in these different groups. However, the Coalition's mailing list that was given to me to send out questionnaires included 1,200 women. The core groups of the Coalition's organizations comprised Jewish and Palestinian women, Israeli citizens, with previous experience in mixed protest movements, and feminist organizations (65%-90%). The questionnaires distributed to activists in a range of organizations show quite consistently that these were women in their mid-forties (mean age=45), of Western European/ American origin (95%), predominantly secular, university-educated (72%-85% depending on the organization), and primarily in the social sciences and the humanities. More than 80% were active in the labor market and the majority practiced professions typical of the new middle class – engineering, medicine, law, psychology, social work, consulting, university lecturing, and teaching (65%).The Coalition had a polymorphous structure and it exhibited great pluralism, being composed of several single-issue, autonomous groups. The groups varied in their activities, political version of left and feminism, and mode of organization. For Example: *Shani-Women Against the Occupation* was established to promote educational activities and dialogue between Jewish and Palestinian women, Israeli citizens, and between Israeli and Palestinian women. *Women in Black* was and still is determined to keep the issue of the occupation on the public agenda and conducts vigils on a regular, recurrent basis. At the height of its activity, there were about 35 vigils in Israel. They stand at central intersections in town centers and at intercity road junctions every Friday afternoon between 13:00 and 14:00 p.m. They wear black clothes, carrying signs with the slogan "stop the occupation". In mixed cities (Haifa, Jerusalem, and Akko), Palestinian Israeli women also participate in this special form of protest. This mode of action – standing in single-sex groups, the black dress, the fixed day and hour, and the fixed slogan – poses a strong challenge to accepted gender definitions and designated traditional roles, and to national boundaries of allegiance (Benski 2005; Benski 2011; Helman 2011). The women reject the maternal principle and are adamant that their protest is based on their citizenship and not on maternal roles and values. They reject maternal discourse and, in their words "refuse to be the womb in the public space." Another group, *Women for Women Political Prisoners* consists of a few scores of women in Jerusalem and Tel-Aviv who lent legal and humanitarian aid to Palestinian women who were imprisoned in Israel during their Intifada activism and kept surveillance of the treatment that these women receive from the prison authorities with special emphasis on violations of human rights including sexual harassment complaints. These two last groups posed the strongest challenge to Israeli society and were considered traitors who switched sides in the conflict.

An analysis of the long list of activities of the Coalition and its constituent groups reveals at least five main anchors around which the activities were organized: 1. the political pressure anchor with protest acts performed mostly by *Women in Black;* 2. pragmatic humanistic action like supplying practical and legal aid to Palestinian women political prisoners and helping to maintain official and secret schools by supplying the necessary equipment such as chalk, notebooks, and pencils (*TANDI*); 3. an educational anchor which was aimed mainly at the empowerment and politicization of women, and at supplying information concerning the predicament of Palestinian women during the uprising *(SHANI, TANDI)*; 4. a symbolic expressive anchor like *Women in Black* vigils and the *Peace Cloh/Quilt* ; and 5. the interpersonal anchor which represents actions aimed at breaking personal barriers and enhancing ties between Jewish Israeli and Palestinian women (the *Bridge to Peace*, and *Bridge*). All of these actions represent different expressions of solidarity and empathy with Palestinian women and their predicament.

Value orientations

The Value orientations of the coalition consist of three interrelated value systems: universal humanistic values, political leftwing values, and feminism. One of the women put forward a rather challenging message: «Our protest is twofold: against the oppression of one people by another, and against gender-based oppression of women by men» (C. S. *address 1989*).

A more complex argument appeared a few years later:

> The occupation is not a phenomenon that exists only on the other side of the border. There is a social, cultural-normative system that has given rise to the occupation and sustains it. The very same system is the system of values and institutions that exploits women and uses the occupation to legitimize the oppression of women in Israel. ... This is a masculine militaristic value system ...; it sanctifies the state and security.... Without fighting the roots of the occupation, which is the Israeli expression of patriarchy, no attempt at achieving equality can succeed. (T.P., May 1993 at the Feminist Seminar at Givat Haviva).

The views expressed in these quotes demonstrate the anti-systemic nature of their claims. The *Coalition* directs a feminist critique at the 'rules of the game' of the institutionalized political system which is perceived as representing 'masculine' rules. They demand the right to autonomy, equality and self-determination for both women and Palestinians. Militaristic culture is mentioned in the last quote, but is not accorded a central position. This would change in the second wave of women's peace activism.

Globalization

Here we see the beginnings of globalization in several ways. *WILPF* was a branch of an international organization; *Women in Black* were inspired by the *Mothers of the Plaza de Mayo* and were the first to break into the global world and inspire many groups of women around the world who adopted the practice of *Women in Black* to protest their local issues and the occupation of the Palestinian territories by Israel. At the height of its mobilization, there were over 150 *Women in Black* vigils across the world.

• *The Women's Peace Net, The Jerusalem-Link and Bat-Shalom*

Like *Women and Peace*, The *Net*, established in June 1989, is an organization of Israeli Jewish and Palestinian women (Israeli citizens). Its participants represent a wide spectrum of political views that are acceptable in the Zionist left wing of the Israeli political scene. Its members include women from existing women's peace groups, on an individual basis, prominent women politicians, artists, and scholars. The *Women's Peace Net* arose from an international conference entitled "Give Peace a Chance: Women Speak Out", that was held in Brussels on May 1989 (yet another result of globalization). In one of the documents summarizing three years of activities, *The Women's Peace Net (Reshet)* (no date, probably 1993), the opening sentences state:

> The Israeli Women's Peace Net is a broad-based coalition of women whose purpose is to advance the cause of a negotiated settlement of the Arab-Israeli conflict based on the principles of self-determination, sovereignty, and mutual security for both Palestinians and Israelis. Its activities are a vehicle for breaking psychological barriers between Israelis and Palestinians, building mutual confidence, and most significantly providing a channel for the assertion of a women's voice in the quest for peace, based on the value of equality, respect for differences and interdependence.

Participants, organization and activities

There were 2,000 women on the mailing list of the *Net* (reported at the meeting of the Forum on 4.12.1992). The women, participants of the *NET* came from the more established institutional strata of society, with left-center political views, and the list includes women Knesset members and women from various political parties of the left. This composition was totally unique within the women's peace camp. These women were part of the established Israeli political system. These parties were part of the opposition up to 1992, when following the rise to power of the left they became part of the ruling coalition. This had an additional moderating effect that at times prevented signatures on declarations and documents that were opposed to the party line.

The *Net*'s structure includes autonomous local branches with similar types of activities and goals. This is different from the Coalition that had a more federal structure of different movements and organizations. The central decision-making body is the Forum which meets once a month to discuss issues and reach decisions. (I participated in and recorded most Forum meetings since 1992 till its merger with Bat-Shalom in 1995). The movement cooperates with other peace movements, including the coalition of *Women and Peace*. This occurs mostly on national-level marches and events such as International Women's Day and International Peace Day.

Compared with the *Coalition*, the repertoire of action includes more educational, awareness raising activities and meetings with Palestinian women's organizations across the line than vigil acts.

The goals that the *Women's Peace Net* seeks to accomplish are consistent with the political views of the moderate elements of the Zionist left. They believe in the right to self-determination of the Palestinian people and in their right to choose their own representatives. They support negotiations with the PLO – the legitimate representatives of the Palestinian people. They support the principle of 'two states for two peoples' and they are opposed to the continued occupation of the territories. However, unlike the *Coalition,* they believe that these goals can be obtained within the institutionalized political system by exerting pressure on the government to accept the right to self-determination and sovereignty of the Palestinians, and to put an end to the occupation. The *Net* also directs its activities against the Jewish settlements in the territories, claiming that they pose a threat to peace.

Value orientations

The analysis of documents and meetings of the Forum of the *NET* show the same value orientation as the Coalition, only on a more moderate level. The left and the feminist identities were more consistent with a more moderate, liberal feminist approach. Their demands in the political sphere are not based on a radical feminist ideology and they do not demand a redefinition of the concepts of femininity and politics. They do not mention the oppression of women. Consequently, they also do not make the equation between the oppression of women and the oppression of the Palestinian people. They do not perceive the established values and the rules of the political game as reflecting male-centered values and do not reject these rules. They believe that women have the capacity to contribute towards peace through breaking the psychological barriers of misunderstanding and distrust. They believe that through personal contact and discourse between Israeli Jewish and Palestinian women they can build an infrastructure based on trust, which will help to overcome difficult and controversial issues. In other words, they do not pose an anti-systemic critique in the same sense that *Women and Peace* do.

Principled cooperation between Israeli and Palestinian Women –
The Jerusalem Link and Bat-Shalom:

A year after the establishment of the Women's Peace Net, a second set of meetings of prominent Jewish Israeli and Palestinian women was held in Brussels. The women felt that it was time to move on from talks with

Palestinian women to join forces in activities across the divide and establish a cooperative project. The EU supplied the grant for the project and *The Jerusalem Link* was formed. It led to the establishment of *Bat-Shalon* on the Israeli side, with an office in West Jerusalem, and a coalition of Palestinian women's organizations (*Jerusalem Center for Women*) with offices in East Jerusalem. It was an alliance based on agreed upon goals, chief among which was to put an end to the Israeli occupation of Palestinian territories. When this project started, one of the initiators of the project said:

> The first thing that we had in mind in September 1992, when we planned this cooperative project, was that we and the Palestinian women will create a model, that is actually the model of how we see the relationships between Israelis and Palestinians, according to our political vision. For this reason, we deliberately set up two independent centers, which are in constant interaction. An Israeli center and a Palestinian center. Each side is independent and at the same time they cooperate all the time. I consider this unique structure to be a very important political statement of how we see things and how things should be done at the national level (10.2.94, N.H.)

Almost simultaneously with the emergence of Bat Shalom in Jerusalem in 1993, the affiliated group of women that called themselves "Bat Shalom of the North" came into being (Cockburn 2014). In 1995, the Net and Bat-Shalom merged and the name was changed to Bat-Shalom-The Women's Peace Net.

6.2.3.2 *The interim period 1996-2000: Four Mothers*

Since *Mothers Against Silence* dispersed in 1985, despite the high toll in lives, little protest was directed at the war in Lebanon or at the northern border of Israel. This changed drastically when on February 1997, two helicopters carrying soldiers to the army stationed in southern Lebanon crashed in the air. 73 soldiers were killed in this accident. Following this heavy loss of life, a new movement, *Four Mothers* (in Hebrew: *Arba Imahot*) was established on 5 February 1997 (Lemish and Barzel 2000). The founding group consisted of four mothers whose sons were serving in the IDF at the time and they were all living in the northern part of Israel near the border with Lebanon. Like the earlier mothers' movement, *Four Mothers* used maternal concern for the lives of their sons to base their claims and did not engage with the occupation. They gained wide support among the public. At a certain stage, men joined the movement and were present at all the meetings and conferences covered by the media.

The movement organized protest acts, meetings with public figures and politicians, conferences with university academics, many media appearances, etc. It was one of the most visible and most successful women's peace organizations in Israel. Many have attributed the withdrawal of the IDF from Lebanon in May 2000 to their campaigns.

During this interim period, several of the organizations of the first wave continued their activities when peace seemed to be on the way, although there was a significant slowdown after the signing of the Oslo accords in 1993.

6.2.3.3 *The second cycle of women's peace activism*

• *The Coalition of Women for Just Peace*

In October 2000, a provocative visit by Ariel Sharon – the then leader of the political opposition in the Israeli Knesset – to the Al-Aqsa mosque in Jerusalem sparked the "El-Aqsa Intifada" or "the second Intifada". It gave rise to a second wave of peace activism. The *Coalition of Women for Just Peace* is the largest new feminist organization against the occupation of Palestine and for a just peace, founded in November 2000, a month after the outbreak of the Second Intifada. It describes itself on its website and various documents as «a feminist movement of activists and a network of feminist and women's organizations against the occupation and for social justice» (https://coalitionofwomen.org/?lang=IW). The *Coalition* demands «a deep and radical change of society – from a militaristic conquering society to an equalitarian and feminist society» (The Coalition work-plan for 2010, internal document.).

The new *Coalition* included individual women and nine feminist organizations, 6 of which were veteran groups *(Women in Black, TANDI – Movement of Democratic Women in Israel,* the Israeli branch of *WILPF – Women's International League for Peace and Freedom, Bat-Shalom, NELED-Women for Coexistence, Noga Feminist Journal),* and three new ones *(Women and Mothers for Peace* (some of the women were formerly members of *Four Mothers* and later founded the *Fifth Mother), New Profile* and *Machsom Watch* (Benski 2006, 2010).

The goals of the Coalition were formulated as nine principles, accepted by all its constituent groups: (1) an end to Israel's occupation of Palestinian territory; (2) women's participation in peace negotiations; (3) the establishment of a Palestinian state alongside an Israeli state, following the 1967 borders; (4) recognition of Jerusalem as the capital common to the two states; (5) Israel's recognition of its responsibility for the outcomes of the 1948 war, which obliged the state to search for a just solution to the Palestinian refugee problem; (6) equality, justice, and cooperation with Israeli Palestinians; (7) rejection of militarism in Israeli society; (8) equal rights for women and for all Israel's residents; and (9) social and economic justice for the State's citizens, together with integration into the Middle East (taken from an internal document). These principles represent the Coalition's sociopolitical platform at its inception; any women's organization that accepts these principles can become a member of this alliance. The demands for equality and a settlement of the Israeli-Palestinian conflict are integral to its comprehensive vision of a just and egalitarian society.

- *Continuities between the first and the second wave of women's peace mobilizations*

The coalition of Women for [just] Peace represents continuity with the first wave *Coalition of Women and Peace* in many respects.

Participants: There are not too many differences between the first and the second wave participants in terms of socio-economic characteristics. In the second wave, my analysis is based on an examination of the details of the participants in two conferences held by the *Coalition* – one in October 2002, the other in December 2002. It appears that the basic profile of the first wave coalition participants remained intact. Most of the participants were Jewish and Palestinian women (Israeli Citizens); the Jewish women were of Ashkenazi origin, middle-aged, predominantly secular, university-educated, primarily in the social sciences and the humanities, and active in the labor market. The main difference lies in the finding that about 25% of the participants in the 2002 conferences were Israeli Palestinian women, most of whom belonged to *TANDI*, the organization that has a large majority of women Palestinian citizens of Israel.

In terms of value orientations, goals, forms of contention and globalization processes, many continuities were also detected, but also some significant processes of change have occurred. Looking at continuities, we can see continuity of the Goals of ending the occupation, promoting peace in the Middle East, demand for Gender Equality and Human rights; continuity of value orientations of the three interrelated value systems of Universal humanistic values, political left wing (peace) values and feminism that we found in the first wave Coalition; continuities in *modus operandi*, flowing from a feminist worldview that upholds decision-making following discussion and the reaching of a consensus rather than top-down imposition of decisions. There is also continuity in the adoption of pluralism and rejection of one sole organizing principle and uniform voice, and a polymorphous structure. Furthermore, the activities organized by the second *Coalition* and its constituent groups can also be classified into the five anchors of the activities of the first coalition.

In terms of Globalization processes, there are continuities and deepening of the processes that started in the first round of activism. We will therefore discuss these when we discuss processes of change and innovation in the second round of activism.

- *Changes between the first and the second wave*

As we stated in the introduction to this chapter, women's peace activism is closely related to the changes and processes of the Israeli-Palestinian conflict. Due to lack of space, I will not specify the dynamism of the occupation, exchanges of fire and special operations of the IDF in the Gaza strip or political changes within Israel. I will only point out that during the years that the women's peace mobilizations have been active, many features of Israeli society and the Israeli occupation have undergone changes. Governing coalitions have emerged and fallen; public attitudes have shifted towards the political right to the extent that the words 'leftist' and 'peacenik' have become almost synonymous with radical betrayal; women's status has changed though not enough to allow them a significant voice in the dominant discourse over security issues; the occupation has become harsher, with many new restrictions and military controls over the civilian population; the erection of barriers, checkpoints and a separation wall between the Palestinian territories and Israel, and the organization strategies, frames and activities of the women have also changed (Herman 2009).

Analytically, three large processes of change that have occurred among the women activists of the new Coalition of Women for Peace can be detected. They are partially overlapping but are different analytically. The first was the maturation and deepening of understanding, analysis, and organization. The second process was the expansion of Goals by adding new goals and diagnostic frames, and the third process was that of the radicalization of action forms, perceptions, and frames.

The first processes that most activists underwent during their activism in the first wave were processes of empowerment and consciousness raising. Nabila Espanioli explains:

> Participation in the activism following the first Intifada and the cooperation between Jewish and Palestinian women had a positive effect on the feminist movement in Israel. Feminist women became more political and the Palestinian women became more feminist... at the start of the process I did not even define myself as feminist. (Nebila Espanioli, in Issachar 2003, pp. 29-30).

Women who came from a left-wing background were exposed to radical feminist critique and human rights values, and vice versa. Their participation in the first coalition, their compliance with common principles and convergence around the three value systems led to the strengthening of each element and their convergence and maturation of identities that are in sharp opposition to hegemonic identity. This led the coalition to anchor their claims and statements more and more on human rights issues which provide the meeting point of these 3 ideological frames.

The second process of the expansion of goals and frames occurred chiefly through the new movements. The chief new frame that turned into an important goal that was added to the *Coalition's* agenda was opposition to militarism and the demilitarizing of Israeli society. All the groups within the coalition share this discourse. Their rhetoric includes phrases such as: "soldiers' state", "army state", "cultural militarism", and "a deep process of militarization". The roots of this identification of a "male militaristic culture" are found in the first wave of activism (see quote on p. 104), but acting against the militaristic culture was adopted *by New Profile* as its main goal, and it was elevated in 'status' and included in the nine principles of the Coalition and became more critical and radical.

New Profile-movement for the Demilitarization of Israeli society was founded in October 1998. The movement presents itself on its website as: "a group of feminist women and men citizens of Israeli society" committed

> to change the Profile of Israeli society from a militarized society of war and might, to an actively peacemaking community in which the rights of all its citizens are protected and promoted equally, and the human rights of all people residing inside and outside Israeli borders are respected (www.newprofile.org).

Its focus on raising public awareness to the level of militarization of Israeli society, deepening critical discussion of this issue, and the assistance that the movement gives to conscientious objectors distinguished *New Profile* from the other social movements in the *Coalition*. The movement views 'cultural militarism' as the major factor shaping Israeli society. They claim that militarism penetrates all walks of life, constructing a 'world taken for granted' at cultural, cognitive, and emotional levels; moreover, the educational system is held responsible for instilling the young with militaristic values.

The third process of change, the radicalization of frames, perceptions, and forms of action is visible first through the list of movements that joined the Coalition. In the first wave, the *Women's Peace Net* and *Bat-Shalom* kept separate from the more radical *Coalition of Woman and Peace*. In the second wave, *Bat-Shalom* joined the second Coalition, which signals the radicalization of the most liberal feminist and institutional section of the first wave. Subsequently, the following principles of the Coalition all represent changes that were and are still considered radical in Israeli society: the principle of the establishment of a Palestinian state; turning Jerusalem into the recognized Capital of the two peoples; the demand that Israel recognizes and accepts responsibility for the outcomes of the 1948 war and looks for a just solution to the Palestinian refugee problem; and the demand to demilitarize Israeli society.

Strong radicalization is also evident in the critique directed by *New Profile* at the basic elements of what they call "cultural militarism". I will mention three of the most radical critiques:

1. They refuse to accept «the assumption that there is a real constant existential threat to the state of Israel» (Mazali 2002). Furthermore, they do not accept the premise that "we have no choice".
2. They question the necessity of the national service law and compulsory army service. No women's movement before or after New-Profile dared to suggest such a radical challenge within Israeli society. They suggest canceling compulsory military service and request that the state should recognize the right of conscientious objection.
3. They attack the gender roles allocated to women in Israel, claiming that enacting these roles represents collaboration with the militarist culture. They «refuse to continue to mobilize, to grow children for the draft, to support partners, brothers, and fathers who do army service, when the heads of state continue to use the army instead of looking for other solutions» (AMANA June 1999).

The second wave also exhibits radicalization in its forms of activities. This is most evident in the actions taken by *Machsom Watch*. *Machsom Watch* – Women for Human Rights, was founded in February 2001. The central activity of *Machsom Watch* consists of witnessing and monitoring the conduct of the soldiers at the military checkpoints, attempting to prevent violations of the human rights of the Palestinians passing through these checkpoints. They compile reports and disseminate them to the public and decision makers. These activities often lead to confrontations and conflicts between the women and soldiers (Resh and Benski 2012). This is the first time that an organized civilian, critical eye has been directed at the conduct of the army, and it has been met by both the army, the soldiers, and the Israeli public with great animosity.

The last process that we will discuss here, is the globalization process where we see both continuity and deepening. Due to the increase and spread of globalization processes, and due to the widespread means of electronic communications, during the second wave living standards facilitated the globalization processes that started in the first wave and opened possibilities that did not exist before. The process that is most obvious in the *Coalition* is the spread of movement ideas, practices, and frames from one country to another. Indeed, processes of diffusion that started in the first wave, continued to spread ideas and inspirations. Furthermore, the second *Coalition* is well integrated in the global information and communication network. This enables the mutual flow of information and of support statements from various international organizations. This has led to an extensive international exposure and recognition of the women's peace organizations that has won them international prizes and nominations for the Nobel Peace prize (*Women in Black* was nominated twice for the Nobel Peace Prize and 8 Israeli women, 4 of whom are activists of the *Coalition* were nominated for the peace Nobel prize by the Swiss 1,000 Women project).

Epilogue

Despite the title of Tamar Herman's book *The Israeli Peace Movement: A Shattered Dream* (2009), Cockburn's paper titled, 'The dialogue that died' (2014), and my own title of former presentations that included the phrase 'emotional alienation of Women's Peace Activists in Israel' (Benski 2011), the story of the women's peace activism in Israel is, by no means done. It is true that movements have dissolved, but others have emerged as the conflict continues, changes phases and faces.

In this paper I wanted to bring the entire story of the women's peace mobilizations in Israel up to the current wave. Due to lack of space I decided to end my paper with the second wave of activism and tried to make the account as broad as possible, and at the same time go into some more details of the processes of change and the creativity and innovative characteristics of these movements. Indeed, the scope of the activism is too wide to cover in one chapter. There are many more aspects to discuss. One of them that I hardly touched upon is the issue of the maternal-vs-citizens' identity in these women's movements which has a strange history indeed. But this is a topic for a totally separate paper.

However, the story continues. *The Coalition of Women for Peace*, which has changed and experienced further widening of the scope, branching out and radicalizing perceptions, principles and value orientations, is presently undergoing processes of dispersion. At the same time, the new movement, *Women Wage Peace*, that was set up immediately following the IDF operation in Gaza in 2014, is more reminiscent of *The Women's Peace Net* than of the *Coalition*. It appears that alongside the more veteran women activists, a new, somewhat younger generation of feminist women emerged, and under the impact of eleven consecutive years of the rule of the right political wing, they have changed the direction of women's peace activism in Israel, its target audience, and as a result also its language and, identity and ways of acting. It is much less confrontational than the second coalition and is more focused on pressuring authorities to enter peace negotiations and involve women in these processes than on ending the occupation. At the same time, there are also movements that are totally dedicated to ending the occupation and their activism is highly confrontational.

Whether this trend of newer generations of activists will revitalize women's peace activism and will continue remains to be seen. We can only hope that the women will continue to persist and persevere, despite deep disappointments and disillusionments.

Bibliographical references

'AMANA June 1999', no author, http://www.newprofile.org/ [accessed 28 January 2004].

Azmon, Y., 'Judaism and the exclusion of Women from The Public Arena', in. Y. Azmon (ed.), *A View into the Lives of Women in Jewish Societies* (Jerusalem: The Zalman Shazar Center for Jewish History, 1995).

— 'War, mothers, and a girl with braids: Involvement of mothers' peace movements in the national discourse in Israel', *Israel Social Science Research*, 12 (1) (1997), 109-128.

— (ed.), *Will You Listen to My Voice? Representations of Women in Israeli Culture* (Tel Aviv: The Van Leer Jerusalem Institute and Hakibbitz Hameuchad Publishing House, 2001).

Ben-Eliezer, U., 'Civil Society and Militarism in Israel: Neo-Militarism and Anti-Militarism in the Post-Hegemonic Era', in M. Al-Haj and U. Ben-Eliezer (eds), *In the name of security: The sociology of peace and war in Israel in changing times* (Haifa: Haifa University Press, 2003), pp. 29-76.

Benski, T., 'The feminization of peace activism in Israel following the intifada' (Presented at the 14th World Congress of the International Sociological Association, Montreal, July 1998).

— 'The voices of women in protest movements' in *Voices and Silences – The position of women in society and politics* (The School of Behavioral Sciences at The College of Management Academic Studies, 2002), pp. 17-23.

— 'Breaching events and the emotional reactions of the public: The case of Women In Black in Israel', in D. King and H. Flam (eds), *Emotions in Social Movements* (Routledge: London, 2005), pp. 57-78.

— 'The coalition of women for a just peace: The feminist peace project as one of the trends to naturalize Israeli society' in U. Cohen, E. Ben-Rafael, A. Bareli and E. Ya'ar (eds), *Israel and modernity: In honor of Moshe Lissak* (Sde Boker, Ben-Gurion University of the Negev; Jerusalem, Yad Izhak Ben-Zvi, 2006), pp. 523-548.

— 'The coalition of women for peace and the civilian alternative in Israeli society', in D. K. Singhroy (ed.), *Dissenting Voices and Transformative Actions: Social Movements in a Globalized World* (New Delhi: Manohar Publications, 2010).

— 'Emotion Maps of Participation in Protest: The Case of Women in Black Against the Occupation in Israel', *Research in Social Movements, Conflict and Change*, 31 (2011), 3-34.

Berkovitch, N., 'Women of Valor: Women and Citizenship in Israel', *Israeli Sociology*, 2 (1) (1999), 277-317.

Bernstein, D., *The Struggle for Equality: Urban Women Workers in PreState Israeli Society* (New York: Praeger, 1987).

Chazan, N., 'Israeli Women and Peace Activism', in B. Swirski and M. P. Safhir (eds.), *Calling the Equality Bluff: Women in Israel* (New York: Pergamon Press, 1991), pp. 152-161.

Cockbutn, C., 'The Dialogue That Died', *International Feminist Journal of Politics,* 16 (3) (2014), 430-447.

Deutsch, Y., 'Israeli Women against the Occupation: Political Growth and Persistence of Ideology', in T. Mayer (ed.), *Women and the Israeli*

Occupation: The Politics of Change (London: Routledge 1994), pp. 78-93.

Deutsch-Nadir, S., 'Capitalizing on women's traditional roles in Israeli peace activism' (unpublished master's thesis, The Fletcher School, Tuft University, Oregon, 2005).

Elior, R., '"Present-Absent", "Still Nature" and "A Beautiful Maiden who has no Eyes": The Question of Presence and Absence of Women in the Holy Language, the Jewish Religion and in Israeli Reality"', in Y. Atzmon (ed.), *Will You Listen to My Voice? Representations of Women in Israeli Culture* (Tel Aviv: The Van Leer Jerusalem Institute and Hakibbitz Hameuchad Publishing House, 2001), pp. 42-82.

Fogiel Bijaoui, S., 'On the Way to Equality? The Struggle for Women's Suffrage in the Jewish Yishuv, 1917–1926', *Megamot, Behavioural Science Quarterly*, 34 (2) (1992), 262-285.

— 'Women in Israel: The Social Construction of Citizenship as NonIssue', *Israel Social Science Research*, 12 (1) (1997), 1-30.

Freeman, J., 'On the Origins of the Women's Liberation Movement', *American Journal of Sociology*, 78 (1973), 792-811.

Gilath, N., 'Women Against War: 'Parents Against Silence'", in B. Swirski and M. P. Safir (eds), *Calling the quality Bluff: Women in Israel* (New York: Pergamon Press 1991).

Helman, S., 'From Soldiering and Motherhood to Citizenship: A Study of Four Israeli Peace Protest Movements', *Social Politics*, 6 (3) (1999), 292-313.

Helman, S., and T. Rapoport, '"These are Ashkenazi women, alone, whores of Arafat, don't believe in God, and don't love Erez Israel": Women in Black and the Challenging of the Social Order', *Theory and Criticism*, 10 (Summer 1997), 175-192.

Herman, S. T., *The Israeli Peace Movement A Shattered Dream* (Cambridge: Cambridge University Press, 2009).

Herzog, H., *Realistic Women: Women in Local Israeli Politics* (Jerusalem: The Jerusalem Institute for Israel Studies, 1994).

Herzog, H., *Gendering politics – women in Israel* (Ann Arbor, MI: University of Michigan Press, 1999).

— 'The Fighting Family: The Impact of the Arab-Israeli Conflict on the Status of Women in Israel', in M. Al-Haj and U. Ben-Eliezer (eds), *In The Name of Security: The Sociology of Peace and War in Israel in Changing Times* (Haifa: Haifa University Press 2003), pp. 401-420.

— 'Re/Visioning the Women's Movement in Israel', *Citizenship Studies*, 12 (3) (2008), 265-282.
Isachar, H., *Sisters in Peace: Feminist Voices of the Left* (Tel-Aviv: Resling Publishing, 2003).
Kimmerling, B., 'Militarism in Israeli Society', *Theory and Criticism*, 4 (1993), 123-140.
Lemish, D., and I. Barzel, 'Four Mothers: The Womb in the Public Sphere', *European Journal of Communication*, 15(2) (2000), 147-169.
Mayer, T. (ed.), *Women and the Israeli Occupation: The Politics of Change* (London: Routledge, 1994).
Mazali, R., 'Women Refuse', http://www.newprofile.org/showdata.asp?pid =252 [accessed 13 April 2002].
Melman, B. (ed.), *Borderlines: Genders and Identities in War and Peace 1870-1930* (London: Routledge, 1998).
Pope, J. J., 'The emergence of a Joint Israeli-Palestinian Women's Peace Movement During the Intifada' in H. Afshar (ed.), *Women in the Middle East: Perceptions, Realities and Struggles* (London: Macmillan Press LTD, 1993) pp. 172-184.
Raday, F., Shalev, C., and M. Liban-Koby (eds), *Women's Status in Israeli Law and Society* (Tel-Aviv: Shocken Publishing House Ltd., 1995).
Reardon, B., 'Sexism and the war mechanism', in H. Gor (ed.), *The Militarization of Education* (Tel Aviv: Babel 2005), pp. 59-85.
Resh, N., and T. Benski, 'Women meet soldiers: An ambivalent encounter', *Journal of Research in Peace, Gender and Development*, 2(13) (2012), 293-303.
Safran, H., 'The American Connection: The Influence of the American Feminism on the struggle for women's Suffrage in the Yishuv (1919-1926) and Women's Equality in Israel (1971-1982)' (Doctoral dissertation, Department of General History, Haifa University, Haifa, 2001)
Sasson-Levi, O., 'Radical Rhetorics, Conformist Practices: Theory and Praxis in an Israeli Protest Movement'. Shaine Working Papers No. 1 (The Hebrew University of Jerusalem, The Social Sciences Faculty, 1995)
Shadmi, E. 'Women in Black: Social Discourse in the Streets'. A paper presented in the conference on *Women and Nationality in Israeli Society* (The Van Leer Jerusalem Institute, December 1993).
Smooha, S., 'Implications of the Transition to Peace for Israeli Society', in A. Majid and U. Ben-Eliezer (eds), *In The Name of Security: The*

Sociology of Peace and War in Israel in Changing Times (Haifa: Haifa University Press, 2003), pp. 455-488.

Svirsky, G., 'Standing for Peace: A History of Women in Black', www.gilasvirsky.com/wib_book.html (1996) [accessed 19 January 2010].

— 'The Women's Peace Movement I Israel', in M. Kaplan, and S. M. Rich *Jewish Feminism in Israel: Some contemporary Perspectives* (Hanover: Brandeis University Press, 2003), pp. 113-131.

Zuckerman-Bareli, C., and T. Benski, 'Parents Against Silence: Preconditions and processes in the emergence of a protest movement', *Megamot*, 32 (1) (1989), 27-42.

ANTIMO LUIGI FARRO, SIMONE MADDANU[*]

7.
POPULAR POPULISM

Introduction

The new millennium is characterized by the emergence of new global powers that impose organized social models which are cut off from most people's everyday experiences (Touraine 2018). The systemic powers that encompass the major forces impacting and affecting structural levels of society can be summarized as follows: a) financial flows, b) global investments, and c) new techno-structures. a) Financial flows – also fed by illegal capital – occur in a manner that is separated from the real economy and the practical work-experience of people. b) The allocation of investments on a global scale increases the disparities between developed and non-developed geographical areas, with direct consequences for the individuals and groups living in under-developed areas and affecting their ability to access vital resources. c) The emergence of new techno-structures has given rise to the application of techno-scientific models that aim at shaping the evolution of individual existence (for instance, in fields such as medicine and foodstuffs). The new systemic powers benefit from weak local, nation, regional, and global political institutions. Meanwhile, societies are witnessing a wave of so-called populism on a global scale. From Europe to the Americas and Asia, a fluid digital participation has nurtured new agendas, political parties, and grassroots movements. In some cases, these agencies experiment with short or long-term executive experiences as new trendy political parties. In other cases, the populist approach penetrates traditional political parties, including those that are historically both left-wing and conservative right-wing. Dismissive of history but cognizant of popular apprehensions, modern populism has produced a new political project, albeit with echoes of the past. In some cases, when populism is in charge it defines new governmental policies,

---

[*] This chapter is a work of shared reflection. Simone Maddanu wrote the paragraphs *Introduction* and *Populisms*. Antimo Luigi Farro wrote *A Global Phenomenon* and *Popular Populism: a sociological hypothesis*.

communication, and rhetoric around the globe (Stockemer 2019; Moffitt 2016). Perhaps most importantly, this popular populism introduces a permissible anti-science discourse into the public sphere, which is anti-ethical and anti-evidence (Bergmann, 2018; Bronner 2013), and a subtle form of discrimination in the name of the People, an imaginary majority that eventually reinforces racism. By refusing scientific certitudes concerning climate change (referred to as a hoax), allowing conspiracy theories about vaccines or the use of specific treatment (including cancer therapies), criticizing the role and expertise of intellectuals versus the so-called common sense of the people, populist leaders encourage a public delegitimization of some social institutions. Based on a common gut-feeling – and for this reason it is considered more "authentic" than the "artificial" logic of the technical society – populism must generally be interpreted as a fundamental reactionary force rising up against the status quo: technocracy, and national and supranational institutions, including economics, politics, media, and science (Otto and Köhler 2018).

In this essay we attempt to identify a specific populism as a cultural and social phenomenon, led by political actors who successfully articulate and popularize an agenda, while considering specific contemporary topics that have arisen in the age of globalization. Immigration, nationalism, and national identity are among the most recurrent issues. A general criticism and opposition to elitism – including political institutions as well as scientific communities or the media – represent the common denominator for all forms of populism. Nevertheless, the form of populism that we define as "popular" incarnates neither a specific policy, nor politics. Popular populism is a posture, a narration, and a political, cultural, and social imaginary.

The analysis of populism in terms of political discourses appears to be more effective in its attempt to comprehend these phenomena: They connect to a mere political explanation, while misconceiving the social and cultural shifts in contemporary societies. In the conclusions, we will address the issues for a contemporary sociological approach in order to introduce a new perspective on the empirical phenomenon of "popular populism" around the world. By defining some common denominators, we will argue that populism is particularly popular today around the world due, on the one hand, to the disconnected layers of society and the relative gaps engendered by economic monopolies, including the concentration of wealth – and power – on a global scale. Distant global capitalist forces of financial flows, global investments, and techno-structures along with the easy access to information and the new "digital democracy" create a climate ripe for populist expression[1]. On the other hand, the extent of digital democracy has created both an enlarged networked body, which considers itself as the true People, and a counter public that challenges the elites as well as the blurry power responsible for the unwanted transformation of society (Norris, Inglehart 2019). From that point, our conclusions will discuss how popular populism is an attractive political, ideological, and cultural space more likely to be welcomed by the right-wing in western countries. In this perspective, the most accomplished populist narrations are those able to gather together people around a sense of belonging, such as nationalism and ethnocentrism, which eventually lead to xenophobia, authoritarianism, and fascism (Eatwell and Goodwin 2018; Kaufmann 2018; Mudde 2005; 2007; 2017). Populism, as we observe it here, fits better to right-wing, radical right, and reactionary movements.

7.1 *Populisms*

As has been observed in examples of populism in the last two centuries, from the Americas to Russia and Eastern Europe (Ionescu, Gellner 1969), idealistic references to "the People" (Canovan 2005) are based on an alleged natural and almost sacred agrarian community, as well as political projects (Canovan 1981; MacRae 1969: 155-156; Hofstadter 1969). In summary, the ideological features of such a traditional populism are based on reactionary stances in defense of a community that is no longer represented in modern society. Populism arose as a protest against a disconnected bureaucratic and elitist system that was accused of cutting off "natural" belongings, the sense of community and its fundamental rights to govern over the elites. We will discuss whether these elements still apply to the populism of today and if so, in which way. Subsequently, we will clarify why we need a sociological notion to better assess "popular populism", as we define it.

In a fundamental book that investigates the twisting concept of populism, Margaret Canovan selected a series of historic movements that could be given the populism label:

Agrarian Populisms
1. farmers' radicalism (e.g., the U.S. People's Party)
2. peasant movements (e.g., the East European Green Rising)
3. intellectual agrarian socialism (e.g., the narodniki)

---

1   See Castells 2015; Gerbaudo 2019; Hacker and van Dijk 2000; Norris 2001.

Political Populisms
4. populist dictatorship (e.g., Peron)
5. populist democracy (i.e., calls for referendums and "participation")
6. reactionary populism (e.g., George Wallace and his followers)
7. politicians' populism (i.e., broad, non-ideological coalition-building that draws on the unificatory appeal of "the people")
(Canovan 1981, 13).

Canovan's theoretical classification of populism marks a major distinction between the agrarian agency and the political stance. Due to the complexity of nominating the ideological features of such a variety of perspectives, the empirical examples labeled as populist appear widely different. As described above, the adjective "populism" characterizing these different phenomena appears to be "accidental" (Canovan 1981: 14). Moving from Wiles' analysis of populism as a "syndrome", Canovan cannot identify a common ideology or socioeconomic condition that encompasses all forms of populism. However, "[a]ll forms of populism without exception involve some kind of exaltation of and appeal to 'the people,' and all are in one sense or another anti-elitist" (Canovan 1981: 295).

Traditionally, the notion of populism applies to a wide range of political phenomena, even those in opposition to each other. The main characteristic of populism lies in a broad idea of People who share a general opposition to a systemic political power, social actors or an authority that constitute – or are perceived as – a "dominant block" (Laclau 2005). Under this umbrella, populism can manifest itself in a variety of associated individuals and groups with different interests and both cultural and political orientations. Viewed in this perspective, populism would include diverse political phenomena of different natures and outcomes. For instance, this definition would apply to small political groups as well as more consistent ones, either confined to the minority or majority in the government; it would also include groups that are anchored in a democratic system or others that cross the democratic threshold, even attempting to overthrow the democratic institutions in order to establish a dictatorship or other authoritarian stance. Thus, this notion of populism could be applied to describe historical phenomena like Fascism and Nazism in Europe and Japan in the first half of the last century. In the meantime, populism would underlie totally different political experiences in South-America (Germani 1978), like Argentina under Juan Peròn during the last century, Hugo Chavez in Venezuela (Ellner, Hellinger 2003) at the beginning of the new millennium, and Evo Morales in Bolivia (Van Cott 2005).

The same notion would include a democratic populism that arose in the last decade of the twentieth century and still operates in Western Europe, characterized by leaders that, on the one hand, accept the legal and formal structure of the democratic institutions and, on the other hand, seek a direct, privileged, and exclusive relationship with the population through media infrastructures and supported by collaborators, and political allies. In order to do so, the populist leaders make extensive use of opinion polls to intercept popular cultural trends and needs, and present themselves as the incarnation of people's desires. Being the people's champion (or captain) means pretending to convey the People's messages and expectations inside the government or among the opposition, through media channels (mainstream or alternative), social media, and social networks. They target a part of the population, using symbols and messages, voluntarily energizing only a part of the population in a divisive way. Through these communicative, manipulative messages, different styles are used to adapt to national or local audiences.

Some scholars agree on categorizing different forms of populism: According to March and Mudde (2005), we can distinguish between right-wing populists, social populists, and neoliberal populists (Mudde 2007: 29-30). Starting from the last one, at first glance, the core of the common definition of populism should conflict with neoliberalism. Where is the criticism of and struggle against elites when it comes to a neoliberal stance? The contradictions still exist, although the new trajectories of some right-wing parties show how neoliberalism is sold as a necessity (Andersen and Bjørklund 2000; Hainsworth 2008: 69, 87-88). A flip-flopping combination of protectionism and liberal market seems to convince the populist base, in some cases renouncing the "nativist" aspect of the right-wing populism (Mudde 2007: 47-48).

Left-wing populism is sometimes considered as a necessary passage to cope with lack of democracy in the neoliberal system. Calling for the power and sovereignty of the People, this approach marks an attempt to capture the populist momentum in our financially globalized societies, in order to renew meaningful conflict against economic elites, and national and global stratification. For instance, Chantal Mouffe suggests a positive reading of populism in the "post-politics", which is characterized by a fuzzy distinction between left and right (Mouffe 2018: 17). By restoring the "democratic ideal" as the power of the people, Mouffe endorses a left populism that could empower the popular sovereignty against the "*oligarchization* of Western European societies" (*Ibid.*: 18). In this perspective, the term sovereignty calls upon the neoliberal system, by opposing the majority of the people to the power of the elites and the current financial capitalism (Kaltwasser 2012; 2017).

The unbridgeable distance between the leading forces of society and people's lives would, then, move some left-wing collective actions to a democratic populism (Gerbaudo 2016) or progressive populism (McKnight 2018), which can be witnessed in the occupation of public squares around the world. These political and cultural movements try to combine an open and inclusive idea of the People with participatory practices, experimenting with new forms of direct democracy. Corporations and multinational companies, financial capitalism, mainstream parties, and political corruption – on a national and global scale – are the main targets of these movements.

While accepting those analyses that underscore the notion of populism as a genuine critique of elitism and oligarchies, we cannot ignore the fact that social populism (Mudde 2007) cannot integrate other features of populism without compromising a progressive agenda, which is based on libertarian stances, defense of minorities, and diversity.

What we observe today in Western Europe and the United States, in addition to Brazil, India, and the Philippines, is a populist rhetoric that becomes popular by dividing the diverse body of society in order to elect a majoritarian core-group. According to Norris and Inglehart, populism happens to be associated with the common notion of authoritarianism, as it is used in political science as well as in social psychology. Consequently, it is quite obvious that the features of the populist agenda encompass some components such as security, conformity, and obedience – loyalty toward the chief who protects the People's interests (Norris Inglehart 2019: 7). Popular populism, as we consider it here, is in part the combination of populism and authoritarianism as a reciprocal legitimating rhetoric. The state of the art on this matter shows an extensive description of the current populism in its right-wing version[2]. These analyses prove how the populist phenomenon has been capturing scholars' attention in the last decades, connecting those radical right-wing tendencies to the historical changes of contemporary societies. Populistic narratives are combined with specific cultural orientations that, in many cases, converge toward an exclusive (in-group) community or a radical right-wing. These populisms easily infiltrate groups and parties that promote right and extreme-right nationalism, as well as rationalistic anti-globalization approaches, or even euro-skeptics

---

2   See Fitzi, Mackert, Turner 2019, Vol 1, 2, 3; Akkerman, de Lange Rooduijn 2016; Betz 1994; Mudde 2007; Hainsworth 2008.

and xenophobics that have been taking the stage in the European Union public sphere in the last decades.

7.2 *A global phenomenon*

In "Populism: Its Meaning and National Characteristics" (Ionescu and Gellner 1969) the authors did not conceive Western European populism, as if a silent anti-elite and "for the People" claims had been intercepted by the Post-WW2 democratic Western institutions. During that time, the economic expansion on the one hand, and organizations such as unions and political parties on the other hand, were at the core of the western European reconstruction, as an improvement in affluent societies.

Since the crisis of the integration system of the European "affluent society", the new liberal order has imposed new models, a new society has emerged, and the financial economy has prevailed over the real economy. Thus, the previous societal references have lost their meanings and led to a consistent destructuring of the previous social life and the construction of a new one (Castells 1998).

Meanwhile, the political structure in western societies has redefined its actors. New political forces – like the Bündnis Zukunft Österreich (BZÖ) of Jörg Haider and the Freiheitliche Partei Österreichs (FPÖ), the British National Party (BNP), and the United Kingdom Independence Party (UKIP) of Farage in the UK, the NF (later RN) of Marine Le Pen in France, the North League of Umberto Bossi in Italy, the Partij voor de Vrijheid (PVV) of Geert Wilders in Holland, and the Vlaams Belang in Belgium – have erupted in the public sphere, eventually entering the national and European parliament (Ignazi, 2003).

The role of the "populist radical right" (Mudde 2007: 26) – a populist form of the radical right – is now a constant in the EU, including the eastern countries of the so-called Visegrád group: the Czech Republic, Hungary, Poland, and Slovakia. An abundant literature has observed the evolution of the right wing discourse in the public space about immigration and Islam in Europe in the last thirty years. To be fair, the populist posture is not the only one that has been framing this debate: Other traditional parties (on both sides) and public figures have raised criticism and concerns about immigration, especially from Muslim countries, and its impact on national identity and cultural transformation. So, what is new?

The attention on populism today is more than justified by an accelerated phenomenon that has involved all western countries in the last decades.

Right-wing parties have arisen in Europe in opposition to immigration, (Betz 1994) amid widespread criticism and phobia – which can be seen as discrimination – regarding the presence of Islam in the public sphere (Akkerman and others 2016: 5; Geisser 2003), and a renewed anti-Semitism. Charismatic political leaders, that we can define as populist, have won over a large number of voters, in some cases reshaping the tradition of the political party – as has been observed in the United States of America (Rahn 2019: 353-357; Kauffman 2018: 30-55; Moffitt 2016: 41-43) or in the very recent ascent of Boris Johnson in the UK (July 2019). In other cases, populism arises as a collective movement looking for decentralizing democracy, swinging between classical left and right, marking a general trend in the political system. These changes appear to be connected to a broader transformation of the social structure – or are even dependent on it.

There are certainly multiple reasons for this revival, but the extent of it clearly shows a global pattern. As we stated at the beginning, the financial flows, global investments, and the new techno-structures are compelling elements of these changes. These forms of populism have become globally popular (Stockemer 2019). It is remarkable that, in recent modern history, humanity has been sharing spaces and knowledge of different cultures on a global scale, remotely as well as closely, like never before. In a broad sense, the complex understanding and direct experience of diversity and cultures have never been so profound and continuous as they are today. Nevertheless, we are facing an increasing and unrestrained popularity of populism which – we can agree now – is successful in its right-wing version, xenophobic and nationalistic. On the contrary, other collective movements aim to construct alternative policies to cope with the consequences of the systemic powers, by advocating a new democracy, tolerance, justice, human dignity, and respect.

7.3 *Popular populism: a sociological hypothesis*

The notion of populism might apply to a huge range of phenomena, even those that are politically opposed to it, that share a common standpoint regarding the idea of the People as a sovereign majority. However, the idea of a popular movement that would embody a popular vision, interests, and political agenda does not seem to assess the political reality around the globe.

If we agree with Canovan when she says that «if the notion of 'populism' did not exist, no social scientist would deliberately invent it" (1981: 301), we have to admit that populism has become popular since mainstream newspapers and media of all kinds started to put on a label on a global political pattern».

The populist phenomenon that we define here cannot be fully understood without considering global social changes: Financial flows, global investments, and new techno-structures are reshaping the structure of the social life on a planetary scale (Touraine 2018).

Popular populism introduces the political construction of specific narratives that are widely imposing counter-platforms, which deal ambiguously with new forms of democracy. Although grassroots agencies and public opinions are clearly recognizable, the populist platform seems to hang on charismatic personalities more than being the result of a collective agency. Even if this populism appears to be a clear effect of a social *disintegration* and decline of the institutions of the traditional system within globalized societies, it does not really address the issue of a redefinition of democracy. Reversing a progressive cultural wave, this popular populism pushes back the political agenda to nationalism and its extents, while keeping interest in the neoliberal system and being profitable for some economic elites. Populism is an approach, a rhetoric, and a narrative that, while promising to unravel the complexity of current societies facing the wide-ranging effects of globalization, constructs an imaginary community of people – sometimes excluding some of them – who look for a simplified society. According to the populist approach, a simplified society cares more about the majority, common sense, collective conscience, and immediacy.

Norris and Inglehart suggest a "backlash" theory based on the analysis of rising values in western societies. According to Inglehart, the post-materialist values engendered a cultural shift, from conservative to socially liberal values. With the statement «they felt they had become strangers in their own land» (Norris and Inglehart 2019: 35), the authors support the thesis of a reaction against the cultural changes in Western societies. Observing this phenomenon from a more general framework, we shall formulate a more drastic sociological hypothesis: popular populism is showing a global paradigmatic change in society. As post-modernity, since Lyotard, has defined a new age characterized by a critique of modernity as a monolithic narrative, in the same way popular populism has arisen in an age of overwhelming disenchantment toward society and its institutions. Nowadays, for a part of the population the disenchantment toward social institutions would include education, media, medicine, science in general, and politics and economics as a perceived whole body detached from individuals. By weakening the legitimacy of these social institutions, the structural social changes introduce a lack of meaning about their social application: For instance, the traditional media are rejected by a part of the population, and a more democratic and networked communication channel is preferred, in which everybody feels legitimized to produce and spread knowledge and information. The brazenness to refute, in a sense, the legitimacy and credibility of each institution lies in the new communication methods, their nature and capabilities. Popular populism is fed by a decentralized communication and participation. Nevertheless, its political representation is more likely to be successful through largely charismatic figures who promise to get rid of moral, ethical, logical, or scientific obstacles to reestablish an imaginary social body.

Bibliographical references

Akkerman, T., de Lange, S. L., and M. Rooduijn, *Radical Right-Wing Populist Parties in Western Europe* (New York: Routledge, 2016).
Andersen, J. G., and T. Bjørklund, 'Radical Right-wing Populism in Scandinavia: from Tax Revolt to Neo-liberalism and Xenophobia', in P. Hainsworth (ed.), *The Politics of the Extreme Right: From the Marginsto the Mainstream* (London: Pinter, 2000), pp. 193-223.
Bergmann, E., *Conspiracy & Populism. The Politics of Misinformation* (Cham: Palgrave Macmillan, 2018).
Betz, H. G., *Radical Right-Wing Populism in Western Europe* (New York: St. Martin's Press, 1994).
Bronner, G., *La Démocratie des crédules* (Paris: PUF, 2013).
Canovan, M., *Populism* (New York: Harcourt Brace Jovanovich, 1981).
— *The Information Age: Economy, Society and the Culture, End of Millennium* (Malden MA: Blackwell, 1998), vol III.
— *The people* (Cambridge: Polity, 2005).
Castells, M., *Networks of outrage and hope: social movements in the internet age* (New York: John Wiley & Sons, 2015).
Eatwell, R., and M. Goodwin, *National Populism: The Revolt Against Liberal Democracy* (Gretna, LA: Pelican 2018).
Fitzi, G., Mackert, J., and B. S. Turner (eds), *Populism and the Crisis of Democracy, vol. 1: Concepts and Theory* (New York, NY: Routledge, 2019).
— *Populism and the Crisis of Democracy, vol. 2: Politics, Social Movements and Extremism* (New York: Routledge, 2019).

— *Populism and the Crisis of Democracy, vol 3: Migration, Gender and Religion* (New York, NY: Routledge, 2019).
Geisser, V., *La Nouvelle islamophobie* (Paris: Editions La Découverte, 2003).
Gerbaudo, P., *The mask and the flag: populism, citizenism, and global protest* (Oxford: Oxford University Press, 2017).
— *The Digital Party: Political Organisation and Online Democracy* (London: Pluto Press, 2019).
Germani, G., *Authoritarianism, Fascism, and National Populism* (New Brunswick: Transaction Books, 1978).
Hacker K. L., and J. van Dijk (eds), *Digital democracy: issues of theory and practice* (London: Sage, 2000).
Hainsworth, P., *The Extreme Right in Western Europe* (New York, NY: Routledge, 2008).
Hofstadter, R., 'North America' in G. Ionescu and E. Gellner (eds), *Populism: Its Meaning and National Characteristics* (London: The Macmillan Company, 1969), pp. 9-27.
Ignazi, P., *Extreme Right Parties in Western Europe* (Oxford & New York, NY: Oxford University Press, 2003).
Ionescu, G., and E. Gellner (eds), *Populism: Its Meaning and National Characteristics* (London: The Macmillan Company, 1969).
Kaltwasser, C. R., 'The Ambivalence of Populism: Threat and Corrective for Democracy,' *Democratization*, 19 (2) (2012), 184-208.
Kaltwasser, C. R., and C. Mudde, *Populism. A Very Short Introduction* (Oxford University Press, 2017).
Kaufmann, E., *Whiteshift. Populism, Immigration and the Future of White Majorities* (UK: Penguin Random House, 2018).
Laclau, E., *Politics and Ideology in Marxist Theory: Capitalism, Fascism, Populism* (London: NLB, 1977).
— *On Popular Reason* (London & New York: Verso, 2005).
March, L., and C. Mudde, 'What's left of the radical left? The European radical left since 1989: decline and mutation', *Comparative European Politics*, 3(1) (2005), 23-49.
McKnight, D., *Populism Now! The Case of Progressive Populism* (Sydney: NewSouth Publishing, University of New South Wales Press, 2018).
Moffitt, B., *The global rise of populism: performance, political style, and representation* (Stanford CA: Stanford University Press, 2016).
Mouffe, C., *For a Left Populism* (London-New York: Verso, 2018).
Mudde, C., *Populist radical right parties in Europe* (New York: Cambridge University Press, 2007).

Norris, P., and R. Inglehard, *Cultural Backlash: Trump, Brexit, and Authoritarian Populism* (Cambridge UK: Cambridge University Press, 2019).

Norris, P., *Digital divide: civic engagement, information poverty, and the internet worldwide* (Cambridge New York: Cambridge University Press, 2001).

Otto, K., and A. Köhler, *Trust in Media and Journalism Empirical Perspectives on Ethics, Norms, Impacts and Populism in Europe* (Wiesbaden: Springer, 2018).

Postel, C., *The Populist Vision* (Oxford: Oxford University Press, 2007).

Rahn, W., 'Populism in the US: The evolution of the Trump constituency', in K. A. Hawkins, R. E. Carlin, L. Littvay and C. R. Kaltwasser (eds), *The Ideational Approach to Populism Concept, Theory, and Analysis* (New York, NY: Routledge, 2019), pp. 350-373.

Stockemer, D., *Populism Around the World: A Comparative Perspective* (Cham: Springer, 2019).

Touraine, A., *Défense de la modernité* (Paris: Seuil, 2018).

ELENA FRASCA

8.
A "NEW MANIA". THE CARBONERIA: CONFLICTS AND AMBIVALENCES (SICILY, 1820-1830)

The establishment of that new state entity called the Kingdom of the Two Sicilies in 1816 marked a process of unification of the territories of Southern Italy that, however, did not correspond to an equal unity of purposes and expectations between the mainland and the island parts of the kingdom. Many people in the regions *al di là del Faro* (beyond the Lighthouse of Messina, i.e. Sicily), keenly felt the lack of the privileges of the old Regnum, the downgrading to a peripheral region and the distance from the court; they were also frustrated by the new administrative system that had become too complex.[1]

The conflict was evident and the underlying ambivalences were undeniable.

The riots of 1820-1821, with the stamp of the *Carboneria* (Romeo 1950; Spagnoletti 1997), were part of a more widespread movement that demonstrated, for some Sicilians, a desire for freedom from Naples, rather than the first step towards Italian unification. The ancient phenomenon of political secret societies arose again in this period and, in an apparently silent and ambiguous way, promised to subvert the established order, drawing an outline of a very different social and sociological profile, of rupture and rebuilding at the same time. However, in the complex administrative mechanism created by the restored Bourbons, based on the Napoleonic reform, we can see sensational instances both of adherence and rejection.

The character of the Sicilian insurrectional movements in 1820-1821 has interested a large number of historians, above all concerning the impact of the process of modernization – based on the French style – on the island, the wisdom of importing the model of Murat also in Sicily (Renda 1968; De Francesco 1992), and the role of the *Carboneria* as a strong social phenomenon (D'Alessandro and Giarrizzo 1989). Therefore, for some

---

1   cf. *Collezione* 1817: 245-290; Manna 1940; Dias 1858; Landi 1977; Iachello 1998: 47-51; De Martino 2000.

Sicilians the idea of separatism from Naples at all costs seemed to be the main reason behind the riots in 1820-1821 (Cortese 1934: 191-214).

Palermo – land of barons and separatist feelings – was the cradle of the revolt because the citizens were aggrieved by the second-rate role to which they had been relegated after the creation of the Kingdom of the Two Sicilies; they firmly believed in the Constitution of 1812 (Sciacca 1963; Laudani 2011). The awareness of having lost some ancient parliamentary prerogatives – the glories of the Regnum and the most recent bicameral experiences in the English style – was at the basis of their discontent, characterized by different feelings about the Sicilian constitution of 1812 and the Spanish one of the same period. The latter, however, had dangerous intentions favouring the Government of Naples. This idea was rejected by Palermo, but not by a large section of the island's east coast (Renda 1968: 42; De Francesco 1996).

Although with the law of December 11, 1820 the king declared he would «preserve the privileges granted by him and his sovereign predecessors to his beloved Sicilians», the parliament was not convened anymore (Cortese 1951: VII), thus nullifying the "Sicilian nation". Besides, in Palermo, the concerns increased with the awareness that the new Bourbon administration (Iachello 1991: 135-138) – based on the Napoleonic one and created in order to make the entire kingdom functional – was an additional obstacle to the libertarian ambitions of towns, crushed by a system which suppressed the courts, offices and free ports, and destroyed ancient legacies and old autonomies (Cingari 1977: 3-83).

Other Sicilian cities – Messina and Catania, in particular – appeared to be more receptive to the new system, probably because they recognized in it a concrete possibility of a trade-off, especially regarding the management of the island's power. The aristocracy's interests were affected (Romeo 1950: 144-147), but, on the contrary, the rising bourgeoisie saw an important opportunity in this new government system, especially in Eastern Sicily where democratic feelings were spreading.

In this framework, we may ask the following questions: what were the social connotations of the revolts of 1820-1821 in Sicily, with its ancient aristocracy, rising democratic feelings and riots? In addition, what was the role of Catania – traditionally closer to the constitutional solutions implemented in the Naples area – during the important events that followed in those turbulent months?

First-hand sources – extrapolated from the documents of the *Intendenza* of the Valley of Catania and from acts by the *Pubblica Sicurezza* (Public Security) and the Police – provide interesting information about secret societies, people suspected of being *Carbonari*, arrests made in the region, and conspiracies against the Bourbon government. In addition, we can also find out about those convicted or detained, and the public spirit of the Valley in the documents. Therefore, the study of Catania's *Intendenza* (Iachello 2010: 175-205), gives us the opportunity to examine a region that responded with some enthusiasm to the political-administrative requests by the Bourbon government, which is evident in the fruitful «*quinquennio riformatore*» (Raffaele 2005: 135-142) (five-year reform) which characterized the first steps of Ferdinand I; they may have seen a way to get some kind of compensation in the new bureaucratic structure.

It is an interesting angle in order to understand the methods of the police and the legal repression put in place in a Sicilian province of undoubted political and social value in a historical period that marks a crucial step in the long and tortuous path toward unification, also in the light of what had been set in motion in 1816 concerning the strong condemnation of secret societies that could become criminal unions. In fact, in that year a law was passed that punished members of illegal companies with five to twenty years' exile: the maximum sentence was given to the founders of secret societies (*Collezione* 1817: 51-52). In early 1821, in order to contain the spread of this phenomenon, decree number 33 (*Collezione* 1821: 8, 56-57) forbade meetings in the countryside with more than five unrelated persons and stipulated the death penalty – by hanging – "or other kind of death suitable for his condition, according to the Penal Code in force" – for the heads, treasurers and directors of the secret society. Ten years' exile and flogging as a public example and a fine awaited those who «knowingly» offered their home or other places for the meetings. Another decree specified that members of an unlawful society, especially the *Carboneria*, and all those who plotted against the state should be declared guilty of high treason and sentenced to death (*Collezione* 1821: 573). Repented members of the "gang" who denounced the names of their members, enjoyed impunity (*Collezione* 1821: 36). After the end of the revolt, a law of September 28, 1822 annulled all the decrees regarding illegal associations published until then, and new ones were enacted. It proceeded to the general amnesty for all those belonging to the banned secret societies (*Collezione* 1822-1825-1828-1831).

In line with what was laid down by the Bourbon laws, this ample documentary source assumes some particularly interesting aspects, allowing us to observe the social and political life in the Valley of Catania during the significant period of the *Carboneria* riots in 1820-1821, and until the first months of the next decade, the 1830s, when anti-Bourbon rebellions broke out. According to Giarrizzo:

it may be that we should see the decade 1821-30, and especially the years after 1824, as a period when *latomismo* – weakened by cultural and political reasons – ossified into some forms of local and parallel power, to which the official power allowed several models of tolerance and cooperation to mature. From this basis, that had not been changed or eroded by the end of the decade, there would emerge without any conflict, some new or renewed experiences of secret societies, traditional plots of organized crime and political "delinquency", resistance that was sometimes invincible to modern and "public" forms of association (D'Alessandro and Giarrizzo 1989: 700).

While driven by forces that were strange and convoluted at times, the *Carboneria* riots in Sicily had the same common goal that would soon find marked characterizations in other forms of revolts.

Alfio Signorelli states that in Catania the government's reaction following these riots was quite soft, and there were many actual or probable instigators of revolutionary events who, after serving short sentences, slotted themselves back into the main roles of urban and provincial administration (Signorelli 2015: 223-225).

In its censorial process, the complex system of inspections provided for the intervention of the various bodies that comprised the supervisory apparatus that the Bourbons wanted, which was also an aspect of the rigid administrative structure established from the first months of the restoration.

From the centre to the suburbs, all the actors called to play a leading role in the urban and regional stage of the Two Sicilies in the first half of the 19th century promptly responded to directives issued from the top: the general director of police, together with his secretary, the lieutenant of Sicily, who was the *alter ego* of the king who lived in Naples, the *intendente*, who was the head of the valleys and therefore represented the extreme propagation of monarchical power in Sicily, are among those who pressed for the investigation of some people considered to be suspects for various reasons.

Among the suspects, the most common accusations were «subject unfaithful to the king; perjury; recidivist *Carbonaro*; intriguing; disruptor; irregular and turbid conduct; accomplice and protector of the *Carbonari*; and dignitary of the *Carboneria*». «Prison; exile; and suspension *a divinis* for clergymen» were the most common sentences. The process was a deep investigation into the suspect's «political and moral» life, a sort of trial outside the court which examined the effective involvement of various characters in the life of the suspect. Therefore, they produced evidence for or against this or that person, accusations against or praise for a man (and the only woman who turned up among the papers) which doubtless represented the key to the most interesting reading of the wide source of information.

Therefore, thanks to the witnesses we can understand the network of relationships between "suspects" and so many others, of different social classes, who are often extremely significant. However, often selfishness and class interests prompted accusers' mouths and pens, anonymous or otherwise. It seems clear that that there were some individual interests among the papers of the police office. Those who exonerated or accused people – who were suspected by government censors for some reason – were always socially significant characters, powerful men who accused or praised someone who was subjected to police control. In particular, they were judicial officials, important local politicians, or men belonging to the professional world.

This is another sign of the mainly "bourgeois" nature of the *Carboneria* revolts that, in Sicily, seemed to take a vague form, characterized by democratic-republican ambitions, desire for freedom and unspecified special interests and – why not? – individual ones (Signorelli 1988; Macry 1988: 47-63).

The apparent dichotomy between general and individual, which has often characterized the parallel ways of sociology and history, seems to find a point of convergence: in the generalizing context of a historical period marked by fracture, sometimes individual paths of people belonging to certain social classes can lead to important choices and only apparently closed circuits.

Among the analyzed social categories, the clergy appears particularly numerous, especially those men belonging to the high Church. As Romeo points out, the close bond between the clergy and the *Carboneria* conspiracies can be explained by the social nature of the island's clergy, principally comprising men from the small intellectual bourgeoisie (Romeo 1950: 160). On the other hand, during his stay in the Caltagirone area on his tour in Sicily to spread the ideas of the *Carboneria* in 1818, the Tuscan poet Bartolomeo Sestini said that clergymen «will do more with the crucifix than they can ever do with the sword» (D'Alessandro and Giarrizzo 1989: 677).

In 1822, there was an outbreak of a repressive phenomenon in Sicily, which also took place in the Valley of Catania. In September of the year before, Pope Pio VII had issued the Bull *Ecclesiam a Jesu Christo* with which he had expressly condemned the *Carboneria* and widened the net of ecclesiastical censure to forms of subversive association – both blatant and hidden – that aimed to shake the spiritual and temporal establishment of the Roman Church.

In 1822 a particularly interesting document was published (ASC *Miscellanea Risorgimentale* 1822: 36-47, 122-124). It was a list of probable *Carbonari* drawn up by the bishop of Caltagirone, Gaetano Maria Trigona, and sent to the magistrate, who sent it to the *intendente* to submit it to the necessary investigations. Those whose names ended up on the «infamous» list were not only clergymen, but also men from other important social classes. In particular, there were several men of the institutions, active personalities in politics who held – or had held – important positions in the management of public affairs, together with numerous other people such as public officers, doctors, lawyers, and so on.

This is an important feature of what happened during this period: it is the sign of a power struggle between the old and new ruling classes, between the old aristocracy, determined to reinvent itself and trying to remain on the top of the power, and a renewed political class that desired to seize the opportunity to climb the social ladder. In the background, however, the feeling of being betrayed by the monarchy persisted – in Caltagirone as elsewhere – among several people who did not accept the fact that Sicily now seemed a peripherical region.

Moreover, a detail that should not be underestimated is the fact that the administrative reform was even more unpalatable for many towns on the island if their names were not included in the division of the Valleys. According to this interpretation, the list of the bishop Trigona becomes more interesting. The ex *sottintendente* Giacomo Maria Aprile Benzo was suspected of «being protector and accomplice of the *Carbonari*», although the superintendent in charge assured the accusers that he had nothing to do with the facts, specifying only that he could «be called weak».

The *primo eletto* (a local council political position), Michele Chiaramonte was accused of being a «recidivist *Carbonaro* before July 1820, and guilty of public employee perjury», while the former mayor Giuseppe Chiarandà was found guilty of being a protector and accomplice of the *Carbonari*.

Onofrio Crescimanno «receiver, public employee» was described as «adventurous»; according to the bishop, he «was assigned to a *Vendita* (group of *Carbonari*), but did not take the oath, as he had done, in the past, in Messina, and for this he had been condemned to death».

«Unfaithful subject to the King» was the accusation against one of the three men belonging to the aristocracy present on the list of the bishop of Caltagirone. This was the knight Emmanuele Bonanno, who had already been subjected to a trial. The baron Martino Caldarera was a fugitive and hiding from the police, while the baronet Rosario Interlandi was «among the *Carbonari* pursued after the events of July 5, 1820».

Among the doctors in the list, Gaetano Fragapane also appears, considered as an «unfaithful subject to the King» who «worked to ruin Caltagirone», as also claimed by the *sottintendente*, while Cosimo Galvano and Gabriele Messina were considered to be members of the temporary *Giunta di Governo*. According to the bishop Trigona, the notary Nicolò Montemagno was a «schemer». A judge from Mineo, Benedetto Carcò, was the founder of the local *Vendita* (group of *Carbonari*), «where he is a speaker».

Several clergymen are transcribed in the list of the bishop Trigona. Among the most interesting cases is that of the parish priest Giacomo Boscari, accused of being an «unfaithful subject to the King», or that of the priest Vincenzo Caruso, considered to be a *Carbonaro* after July 5 byboth the bishop and the *sottintendente*, and for this reason was «suspended *a divinis*».[2] However, the bishop himself was suspected of being «a *Carbonaro*, protector of the *Carbonari* and an enemy of the realists». The *sottintendente* had his answer ready: he did not belong to the *Carboneria* and he did not protect the *Carbonari;* rather, he persecuted all the *Carbonari* priests, indeed, they are still suspended. He enjoys an excellent reputation. He provided information to the government.

Police checks against clergymen continued in the following years throughout the Valley, despite the repression of the riots. From 1823 the *intendente* had the obligation to supervise the clergymen belonging to the diocese of the Valley, in particular those who had been dismissed on charges of belonging to the *Carboneria*, and who were later rehabilitated. A chaplain of the church of Cifali, Santo Rapisarda, was accused of being the «Grand Master of the *Vendita Carbonara* called the *Figli di Pallade*, composed of people who had committed the most brutal crimes» (ASC *Miscellanea Risorgimentale* 1823, 1824: 52-53, 56-66, 69-70).

According to the temporary *sottintendente* of Caltagirone, Baron Sturzo, the high priest of Palagonia Vincenzo Cagenga was a *Carbonaro* «out of necessity», while the priest of Licodia Pietro Sammartino was a *Carbonaro* «for fear of losing his life» and the priest Santo Sampieri «abhorred this pestiferous Institution» (ASC *Miscellanea Risorgimentale* 1821, 1825: 113-115, 242-243). So, the clockwork justice continued to control the work of regular and secular clergymen even later. Following the source,

---

[2] The same suspension fell on the priests Antonino Di Carlo, Giacomo Giaquinta, Vincenzo Libertino, Antonino Lo Monaco, Ferdinando Mariano, Pasquale Montalto, Biagio Morales, Domenico Nicastro, Salvatore Palazzo, Ferdinando Platania, Francesco Randazzini.

however, we note how, especially in the three years from 1825 to 1827, almost all the accusations made against the priests came to nothing.

In March 1825, with the constitution *Quo Graviora*, Pope Leone XII, reiterated the condemnation of any form of subversive association that would operate «to the detriment of the Church and State power». Therefore, clergymen continued to be investigated also over the next years, and so did people with other professions.

An interesting character is Domenico Abramo, a lawyer from Catania, victim of an anonymous accuser who, in a letter addressed to the lieutenant of Sicily in the decisive year 1822, describes him as

> an activist belonging to the *Carboneria*, a man of depraved morals, a disruptive element who gathers these proclamations that the General Rosaroll had spread in the hope of enlisting new forces to the new mania (ASC *Miscellanea Risorgimentale* 1822: 28).

According to the *intendente*, the police inspector Francesco Cannizzaro Grasso was *Great Master* of a sect called *L'Aurora*, in 1822 (ASC *Miscellanea Risorgimentale* 1822: 371).[3] Graziano Cirino, a judge from Nicosia, was a suspect «*Carboneria* orator» in Regalbuto in 1824 (ASC *Miscellanea Risorgimentale* 1824: 116-120). and, two years earlier, he had been accused by the *sottintendente* Aprile Benzo and the bishop Avarna of recommending some of his friends for jobs, for example, the chancellor Mariano Gentile (ASC *Miscellanea Risorgimentale* 1822: 54, 78-79, 116-120). The line between personal interests and hidden hate seems to be very fine.

Several years after the end of the riots, a lawyer from S. Giovanni La Punta, Luigi La Rosa, was accused by his fellow citizens of holding secret meetings at his home and «afflicting the population». However, the judge Milazzo and the inspector of Catania defended La Rosa, specifying that «since La Rosa is a lawyer, he often receives in his own home people who need his help» (ASC *Miscellanea Risorgimentale* 1828, 156-161, 260-265).

The Bourbon police also investigated a small number of doctors suspected of being members of the *Carboneria* and activists in the diffusion of subversive ideas. In 1822, for example, the bishop Avarna and the *sottintendente* accused a doctor from Nicosia, Diego Bonelli Bonelli, of having «a democratic nature» (ASC *Miscellanea Risorgimentale* 1822:

---

3 Another police inspector, Luigi Crisafulli, was accused of being a member of the Carboneria in this year.

265-266). «If he was a member of the *Carboneria*, he was so to avoid violence and mistreatment», said a judge and the mayor about the doctors Salvadore Bellia and Salvadore Cosa, from Paternò, investigated by the police in 1829 and 1830 (ASC *Miscellanea Risorgimentale* 1829-1830: 425-429).

Some aristocrats, as we have seen, appeared on the list of 1822 compiled by the bishop of Caltagirone. But even in the years after the police investigated the noblemen, some of them were accused of dangerous crimes. During the riots, in 1821, Giovanni Bonanno Speciale, baronet of S. Andrea, and the marquis mayor Giuseppe Nicosia, were among the rioters accused of taking part in the revolt of July 23 in Nicosia and setting up a *vendita carbonara* (ASC *Miscellanea Risorgimentale* 1821: 267-300).

The following year, the police arrested count Salvatore Bonsignore, from Leonforte, on the charge of being «in possession of weapons prohibited by the decree of September 11, 1821». In particular, «the Commander of the IV company of the *Gendarmeria* has found rifles, a sabre and two old guns in his home» (ASC *Miscellanea Risorgimentale* 1822: 247). The limitation of the indiscriminate use of weapons was, particularly in those years, the focus of the legislator who, in 1821, promulgated a law about the death penalty for those who «had taken away forbidden weapons» (Frasca 2011: 245-265).

Many Sicilian noblemen took part in the riots, probably to defend their privileges and freedom from the administrative system that the almost ten-year reversal of feudality had not completely removed. The Sicilian aristocracy continued to plot secretly, as can be seen from some documents relating to the years between 1829 and 1831 in which the word «*Carboneria*» still appears with a certain regularity.

In addition to the social *élites*, the police checks also concerned people with different jobs, such as nine «*mastri*» (master craftsmen), two teachers, two tailors, two landowners, a sailor, a shepherd, an engineer and a «*bracciale*» (agricultural worker), a twenty-eight-year old from Pietraperzia, called Angelo Giugno. The document reported his physical appearance: «right body size, gross nose, brown hair, round chin, natural complexion, scar on the right eyelid». For the young man, «accused of taking part in the upheavals of 1820», the general secretary Fogliani asked for his arrest three years later (ASC *Miscellanea Risorgimentale* 1823: 231).

The engineer was Giuseppe Toscano, from Paternò, accused in a lawsuit lodged by the inhabitants of Licodia in 1830 «of carrying out secret meetings and bringing disorder in the towns of Paternò, Biancavilla and Adrano».

According to Anfuso, the judge of Paternò charged with carrying out the investigations, «his conduct was cloudy with regard to his duties as an engineer» (ASC *Miscellanea Risorgimentale* 1830: 409-413). Once again, political accusations and personal hate would seem to blend perfectly.

Finally, the documents reported some people without any information about their status or job. Some of them are very interesting, such as Guglielmo Faillo, from Licodia, who, in 1823, hid in his own home «an illegal wrench and a forbidden knife with some emblems of the *Carboneria*: an axe, a shovel and a hoe» (ASC *Miscellanea Risorgimentale* 1823: 29). The echo of the *Carboneria*, therefore, continued to resonate for a long time among some districts of the island.

The documents analyzed made it possible to observe closely the protagonists and events of one of the most turbulent periods of the 19th century in Sicily. They provided many other suggestions and historical reflections. It is an interesting key to understand how clear and unclear conflicts made for socio-political change in those tumultuous and important years, years that were marked by intense ambivalences that strongly characterized those «secret sociability drawing rooms» that resisted the repressive attacks of the police by the Bourbon censorship, transforming their appearance until the inevitable convergence towards Unification.

The constitutional experience, the riots of 1820-1821 and the following attempts at repression are some stages that marked the way towards that irreversible process which would soon lead to a new and important historical change.

Bibliographical references

Cingari, G., 'Gli ultimi Borboni', in R. Romeo, *Storia della Sicilia* (Napoli: Società Editrice Storia di Napoli, del Mezzogiorno Continentale e della Sicilia, 1977), vol. VIII, pp. 3-83.
Cortese, N., 'Il governo napoletano e la rivoluzione siciliana del 1820-21', *Archivio Storico Messinese*, 35 (1934), 191-214.
— *La prima rivoluzione separatista siciliana (1820-1821)* (Napoli: Libreria Scientifica Editrice, 1951), p. VII.
D'Alessandro, V., and G. Giarrizzo, *La Sicilia dal Vespro all'Unità d'Italia* (Torino: Utet, 1989).
De Francesco, A., *La guerra di Sicilia. Il distretto di Caltagirone nella rivoluzione del 1820-1821* (Acireale: Bonanno, 1992).

— *Rivoluzione e costituzione. Saggi sul democratismo politico nell'Italia napoleonica, 1796-1821* (Napoli: Esi, 1996).
De Martino, A., *Amministrazione e società nel Mezzogiorno del primo Ottocento* (Napoli: Jovene, 2000).
Dias, F., *Introduzione a amministrazione finanziera del Regno delle Due Sicilie* (Napoli: Pellizzone, 1858).
Frasca, E., 'Misfatti e malfattori. Il sistema penale nel Meridione borbonico', in F. Biondi (ed.), *Pensiero politico e istituzioni nella transizione dal Regno borbonico all'Unità d'Italia* (Acireale – Roma: Bonanno, 2011), pp. 245-265.
Iachello, E., 'Appunti sull'amministrazione locale in Sicilia tra la costituzione del 1812 e la riforma amministrativa del 1817', *Rivista Italiana di Studi Napoleonici*, XXVIII (1991), 135-138.
— 'Borbone e stato in Sicilia: la riforma amministrativa del 1817', in Id. (ed.), *I Borbone in Sicilia (1734-1860)* (Catania: Maimone, 1998), pp. 47-51.
— 'Catania nella prima metà dell'Ottocento: poteri e pratiche dello spazio urbano', in Id. (ed.), *Catania. La grande Catania. La nobiltà virtuosa, la borghesia operosa* (Catania: Domenico Sanfilippo Editore, 2010), pp. 175-205.
Labate, V., *Un decennio di carboneria in Sicilia (1821-1831)* (Roma – Milano: Soc. Ed. Dante Alighieri, 1904).
Landi, G., *Istituzioni di diritto pubblico del Regno delle Due Sicilie (1815-1861)* (Milano: Giuffrè, 1977).
Laudani, C., *L'appello dei siciliani alla nazione inglese. Costituzione e costituzionalismo in Sicilia* (Acireale – Roma: Bonanno, 2011).
Macry, P., 'Le ricerche su borghesie e ceti medi nella recente storiografia', in A. Signorelli (ed.), *Le borghesie dell'Ottocento. Fonti, metodi e modelli per una storia sociale delle élites* (Messina: Sicania, 1988), pp. 47-63.
Manna, G., *Il diritto amministrativo del Regno delle Due Sicilie. Saggio teorico storico e positivo* (Napoli: Porcelli, 1840).
Raffaele, S., *La bottega dei saperi. Politica scolastica, percorsi formativi, dinamiche sociali nel Meridione borbonico* (Acireale – Roma: Bonanno, 2005).
Renda, F., *Risorgimento e classi popolari in Sicilia (1820-1821)* (Milano: Feltrinelli, 1968).
Romeo, R., *Il Risorgimento in Sicilia* (Bari: Laterza, 1950).
Sciacca, E., *Riflessi del costituzionalismo europeo in Sicilia (1812-1815)* (Catania: Bonanno, 1963).

Signorelli, A. (ed.), *Le borghesie dell'Ottocento. Fonti, metodi e modelli per una storia sociale delle élites* (Messina: Sicania, 1988).
— *Catania borghese nell'età del Risorgimento. A teatro, al circolo, alle urne* (Milano: FrancoAngeli, 2015), pp. 223-225.
Spagnoletti, A., *Storia del Regno delle Due Sicilie* (Bologna: il Mulino, 1997).

Legislation

Collezione delle leggi e de' decreti reali del Regno delle Due Sicilie:

– decree n. 932, October 11, 1817, *Decreto sull'amministrazione civile de' dominj oltre il Faro*, pp. 245-290.
– a. 1817, f. n. 111, decree n. 789, pp. 51-52.
– 1821, f. n. 4, decree n. 33, pp. 8, 56-57.
– 1821, decree n. 284, p. 573.
– 1821, f. n. 3, decree n. 22, p. 36.
– 1822, f. n. 53, decree n. 670.
– a. 1825, f. n. 4, decree n. 69.
– a. 1825, f. n. 11, decree n. 235.
– a. 1828, f. n. 84, decree n. 1809.
– a. 1831, f. n. 13, decrees nn. 399-404.
– a. 1831, f. n. 15, decrees nn. 424-428.

Archive

Archivio di Stato di Catania (ASC), *Miscellanea Risorgimentale*:

– b. 25, f. I, Palermo, May 21, 1821, Catania, September 27, 1821, cc. 267-300.
– b. 25, f. I, Palermo, July 18, 1821, Catania, December 22, 1825, cc. 113-115, 242-243.
– b. 25, f. I, Palermo, February 4, 1822, cc. 54, 78-79, 116-120.
– b. 25, f. I, Catania, May 9, 1822, c. 371.
– b. 25, f. I, Nicosia, June 17, 1822, June 18, 1822, cc. 265-266.
– b. 25, f. I, Catania, September 28, 1822, c. 247.
– b. 25, f. I, Palermo, February 13, 1823, c. 29.

– b. 25, f. I, Palermo, March 12, 1823, July 3, 1824, cc. 52-53, 56-66, 69-70.
– b. 26, f. III, Palermo, November 26, 1829, January 18, 1830, cc. 425-429.
– b. 26, f. III, Palermo, August 9, 1830, Paternò, November 7, 1830, cc. 409-413.

Sergio Severino, Giada Cascino

9.
Civil Society Versus Institutions: The Historical-Sociological Understanding of a Social Movement Based on Non-Violent Extra-Institutional Tactics

Introduction

At the risk of shocking sociologists, I should be inclined to say that it is their job to render social or historical content more intelligible than it was in the experience of those who lived it. All sociology is a reconstruction that aspires to confer intelligibility on human existences which, like all human existences, are confused and obscure. [...]. For I should even say that the purpose of sociology is to make intelligible what was not so – to reveal the meaning of what was lived without its meaning being perceived by those who lived it. (Aron 1967: 207).

We wanted to *entrust* the beginning of our chapter to Aron's words because they encapsulate the essence of the relationship between sociology and history: in fact, the unveiling of the meaning of social and historical matter gives substance to the axiom «history without sociology is blind; sociology without history is empty» (Topitsch 1975: 110), declining the relationship between the two disciplines according to a paradigmatic synallagma.

However, the identification of the link of reciprocity between sociology and history refers to Weber's thought and his methodological indications, becoming embodied in the method of causal analysis: as Cavalli says, «It is known that, for Weber, there is no scientific knowledge outside the identification of causal links. To explain scientifically always means to explain causally» (1980: 580). Indeed, in *Gesammelte Aufsätze zur Wissenschaftslehre* Weber affirms that history can deviate from "a mere chronicle of events and personalities worthy of memory", adopting a "problematic regulation" that questions itself on a «[...] decisive aspect for the historical elaboration of reality» (1922, It. trans. 1958: 208), be it a person or an event. To this regard, in a reflection on the relationship between historical knowledge and sociology in Weber, Cavalli states that in the classical author's thought, a sociological conceptual framework represents «the main tool to transform historical discourse from narrative-descriptive into problematic-explanatory» (1980: 590).

Reconstructing the Weberian causal relationship, Aron (1967) distinguishes between historical causality and sociological causality, entrusting to the former the determination of the "unique circumstances that have given rise to a given event", and to the latter the identification of a "regular relationship between two phenomena", in which the former represents the favorable or impeding condition of the latter.

According to this reconstruction, the two types of causality are connected by the concept of probability: «[...] the causality between a situation and an event is adequate when we feel that the situation made the event, if not inevitable, at least very probable» (Ibidem: 204). In particular, sociological causality refers to partial and probable relationships, since «[...]a given fragment of reality makes probable or improbable, is favourable or unfavourable to, another fragment of reality» (Ibid: 205).

Aron's analysis of Weberian thought traces what he calls "the interdependence of history and sociology" in the concept of the *ideal type*, that is, a conceptual framework «in which, using the category of objective possibilities, we build connections that our imagination, oriented and disciplined in view of reality, deems adequate» (Weber 1922, It. trans. 1958: 112) and which represents «a partial comprehension of a total whole» (Aron 1967: 207). One of the ideal-types described by Weber is that of *historical particulars* which – in the words of Aron – refer to «a partial reconstruction, since the sociologist selects a certain number of traits from the historical whole to constitute an intelligible entity» (Ibid: 208-209).

Given this interdependence between the two disciplines, sociological causal research – by extension – can be based on the logical procedure of Weber's historical causal analysis which involves four phases:

– construction of the characteristics of the historical particular that one wants to explain;
– analysis of its antecedents;
– unreal modification of one of the antecedents;
– comparison of the unreal becoming with the real evolution.

From the concept of ideal type, as well as from the problematic approach of causal analysis, we have inherited what Weber defines *the judgement of imputation* which we adopted during the research (1922; It. trans. 1958), and which can be summarized in these terms: *selecting and accentuating certain characteristics of a social reality and orienting and disciplining our imagination in view of reality.*

We wonder, however, which characteristics of social reality can (or should) the sociologist select? In what way can the sociologist orient and regulate his imagination in view of reality?

We can find the answer to our questions in that "quality of mind" described by C.W. Mills as "sociological imagination" (1959), a quality that enables us to focus – in the systematic analysis of a social reality – on the intersection between individual biography and history within a given society. Mills himself identifies the possible intelligibility of a social reality in the necessary analysis of these elements, because:

> [...] men do not usually define the troubles they endure in terms of historical changes or institutional contradiction. The well-being they enjoy, they do not usually impute to the big ups and downs of the societies in which they live. Seldom aware of the intricate relationships between the patterns of their own lives and the course of world history, ordinary men do not usually know what this connections means for the kinds of men they are becoming and for the kinds of history-making in which they might take part. They do not possess the quality of mind essential to grasp the interplay of man and society, of biography and history, of self and the world. They cannot cope with their personal troubles in such ways as to control the structural transformations that usually lie behind them (Ibidem: 3-4).

For Mills, the intersection between the elements of sociological imagination can only be grasped by posing "three sorts of questions" (Ibid: 6), which, for the purposes of this work, we summarize as follows:

- identifying the social order of the social reality examined;
- identifying the level of development that the identified social order shows with regard to the evolving of human history;
- outlining the connection between this social order and the historical period examined;
- describing the types of men and women who prevail in this social order.

Moreover, for the author, the essential tool of sociological imagination is the distinction between "the personal troubles of milieu" and "the public issues of social structure" (Ibid: 8): the former refer to the character of the individual and his interaction with the immediate environment; the latter to the overlapping and interpenetration of individual environments in institutions, generating the structure of social and historical life. Difficulties and problems originate from a state of *crisis* determined by the threat to a set of cherished *values* – respectively – for the individual and for a group of individuals.¹

We anchor the possibility of understanding social conflict, with the aim of revealing «[...] the meaning of what was lived without its meaning being perceived by those who lived it» (Aron 1967: 207), to the methodological indications of Weber and Mills: identifying partial and probable relationships between fragments of reality, as well as systematically analyzing biography, history, and social structure.

Therefore, in this chapter, oriented by these indications, we set ourselves the objective of reconstructing the work of unveiling the meaning of a social and historical reality that one of the authors of this contribution carried out in the essay *Intervista impossibile a Danilo Dolci: Saggio sulle funzioni della radio per lo sviluppo dei fatti sociali* – Impossible interview with Danilo Dolci: Essay on the functions of the radio for the development of social facts (Severino 2015): the essay focuses on social conflict that took place in the form of a social movement², in Sicily between the 1950s and the 1970s.

To this end, this reconstruction work is divided into two parts: the first presents a review of the essay, in which we highlight the mutual relationship between biographies and historical events in the social structure, as a substratum of the social movement analyzed; the second proposes the historical-sociological understanding of the social movement under examination, tracing its partial and probable relationships with the factors that favored it.

The work concludes with a reflection on the role of the social movement under consideration in promoting social change.

9.1 *Intervista impossibile a Danilo Dolci: Saggio sulle funzioni della radio per lo sviluppo dei fatti sociali (Impossible interview with Danilo Dolci: Essay on the functions of radio for the development of social facts)*: an analysis of the intersection between biography and history within a society

Intervista impossibile a Danilo Dolci (Impossible interview with Danilo Dolci) is the homodiegetic narration of an "imaginary posthumous

---

1   In addition to the state of crisis, the author includes total dismay, *panic* (threat to all values), *indifference* (absence of preferred values and a threat) and *uneasiness* (absence of a predilection for values and the presence of a feeling of threat that is difficult to define).
2   The social movement is a form of conflict between ordinary people – who have limited power – and those who exercise power (Croteau and Hoynes 2018).

interview" that the author has with Dolci, a man who went down in history as an activist and scholar of the last century, operating throughout Italy, but especially in Sicily.

Focusing the interview on Dolci's civic commitment, the author develops a reflection that aims to analyze the role of the radio in the development of social facts, since «Danilo Dolci's use of radio technologies [...] reflects an intentionality and an awareness that arise from a deep reflection on the political and social role of the media» (Gubitosa 2005). To this end, Severino systematically analyzes the "public problem of social structure" that affected the Valle del Belìce[3], an already impoverished Sicilian region that was torn apart by the 1968 earthquake. The essay highlights how this natural catastrophe amplified the sense of the *threat* to the *cherished values* of the affected communities, and became the emblem of a social catastrophe.

Unemployment, malnutrition and starvation, the logistical incapacity of the State in response to Mafia infiltration into public administrations, and post-earthquake reconstruction of "non-places"[4], which were alien to those regions and that culture, and decisions imposed from above threatened human dignity, honesty and a sense of belonging to the homeland, and generated a state of crisis, the manifest expression of which was a social movement which saw "civil society" and "economic and political institutions", the will of the people, and the unity of the government opposing each other.

As De Nardis points out in the Preface to the essay (2015), Severino's interview represents "a real condemnation of the political and social void" of the social reality analyzed which, although not very extensive in temporal and geographical terms, conveys a representation of our People and our Nation, that is not so distant in time and space.

Precisely this condemnation and this representation – constructed by weaving a plot between the historiography of that social catastrophe and the concepts of classical and contemporary sociological thought – fix the mutual relationship between biography and history in a given society. The author responds to the three sorts of questions highlighted in the *Introduction*, identifying the *public issue of social structure* in the Valle del Belìce:

---

3   La Valle del Belìce is an area in western Sicily between Palermo, Trapani and Agrigento.
4   Places without identity, social relationships or history (Augé 1992; 2005).

- a social order typical of pre-modern societies;
- a connection between the identified pre-modern order and *la questione meridionale* – the cultural and economic divide between north and south of Italy;
- a "human nature" revealed by men and women (and children) repressed, but made aware of (and involved in) change.

The localization and spatial organization of Dolci's civic commitment in Sicily – described by Severino as the activist's anthropic (or human) geography – reveal places that were not affected by the intense social change recorded between the 16th and 19th centuries in European societies: rural areas, villages and towns, poor and dilapidated neighborhoods, and slums[5]; in fact, places appear where the rural dimension, the simple urban planning organization, the agricultural and fishing economy, the economic and technological underdevelopment, the cultural deprivation, and the scant investment in infrastructure did not give way to industrialization, the economic boom after World War II, and urbanization – in short – to modernization, let alone late modernity[6]. These places comprise one of the most miserable and forgotten lands in the south of Italy (Barone 2004).

This social reality reflects *la questione meridionale* – the cultural and economic divide between north and south of Italy, and its implications; from 1861 to the present day, in the south of Italy there has been a lower level of economic development, a different and more backward system of social relationships, and a weaker development of many important aspects of civil life compared to the central-northern regions (Pescosolido 2010). As noted by Pescosolido, the gap between "a persistently agricultural-commercial Southern Italy" and "a noticeably industrialized Northern Italy" widened from the 1880s, reaching its peak at the end of the 1940s. This was aggravated by a persistent "hypostatization of the southern identity" (Severino 2016). Over the years, even if it has never been completely resolved, *la questione meridionale* – the cultural and economic divide between north and south – has, however, prompted the search for possible remedies, aimed even at the political equilibrium of the country, with an overly-optimistic vision of the relative effect on the entire national economy. Just after the Second

---

5 *U puzzu lavari, U scaru, Spiriti Santi, Trazzera vecchia, Cortile Cascino, Borgo Parrini, Mirto, Diga dello Jato* are some of the Sicilian places of the human geography of Dolci.

6 According to Giddens, late modernity is a phase of radicalization and universalization of modernity, an era marked by elements of discontinuity, high pace and global scope of change, and a new nature of institutions (1990, It. trans. 1994).

World War (the historical period in question here) there was «a vigorous resumption of actions of condemnation and proposal by the most important southerners, as well as the parties» (Pescosolido 2010). The actions of Danilo Dolci can be included among these.

The delay in change in the economic sphere represented the breeding ground in which to plant idealistic transformations among the common people, affecting the domain of ideas and values in order to arouse and/or demand transformations in the institutional economic and political spheres: Dolci's repertory of protests[7], expressed by the non-violent struggle – *sciopero alla rovescia*[8] (reverse strike), fasting, counter-information, media activism, demonstrations, graffiti – represents the engine of an idealistic transformation, aimed at spurring economic and political transformations.

It is precisely Dolci's non-violent struggle as a seed of transformation that allows us to identify – in the social and historical reality under examination – the characteristics of a pre-modern society, which we can describe using the dichotomous models of Durkheim and Tönnies.

Those protest actions condemn the communality and sharing of the same conditions among the various social units: the conditions of the *poor people* – unemployed, malnourished, earthquake victims, and children who looked like beggars and idiots – brought to the fore by Dolci as repressed but, at the same time, as individuals who had become aware of (and involved in) change. The similarity and equality of the social units refer, therefore, to a social cohesion based on "mechanical solidarity" (Durkheim 1893; It. trans. 1962)[9], typical of a pre-modern society in which «there is no room for individuality and differences» (Bagnasco, Barbagli and Cavalli 2012: 76). Recalling Tönnies (1887; It. trans. 1963), the communality and sharing of these conditions would refer to a social order describable in terms of *community*, rather than *society*[10].

---

7   The set of means used or available to a group of actors for the claims and for the expression of the identity of a social movement (Tilly 2004).
8   Contrary to the strike, which indicates "an abstention from work by a group of employees" (Bagnasco, Barbagli and Cavalli 2012: 492), the reverse strike indicates the commitment of unemployed people to do a job in order to claim the constitutional right and duty to work.
9   To this, Durkheim contrasts the "organic solidarity" of modern societies, based on the division of labor and the differences between individuals and groups, because they are linked by a relationship of interdependence.
10 Human beings "in the community [...] remain essentially united" in family, neighborly and friendly relationships through elements of communality and sharing, which do not admit extreme levels of inequality; "in society they remain essentially

The repressive sanctions applied to punish Dolci's protest actions reveal – according to Durkheim – a legal system based on criminal law in which there is the submission of social units «to the higher-ranking unit of which they are part» (Bagnasco, Barbagli and Cavalli 2012: 76): in this case, civil society subjected to the regional, and (perhaps even) national, decision-making system corrupted by the Mafia, through «strong pressure exerted on individuals by the collective conscience» (Cascavilla 2018: 278).

Therefore, the historiography of this social catastrophe seems to highlight how the order of this social reality is marked by the assumption of conformist conduct to preserve the *status quo*.

In the contrast between the conservation of the *status quo* and the drive for change recorded in those years, Severino identifies the contribution of the social reality analyzed in the construction of national history. As De Nardis states, the challenge posed to the constituted social order, through actions of non-violent struggle, expresses the path traced in the political and social history of Italy in that period:

> His [Dolci's] teaching is fully in line with the most recent developments in democratic theory, grasping the crux of the current crisis of liberal democracy: the concept of sovereignty (2015: 15-16).

That sovereignty still sees "a disconnection" in the functions of representation between the will of the people and the unity of government (Ibidem). Complaints from eminent politicians and religious leaders, messages of esteem and solidarity from scholars of social reality, and criticism of the acceptance of the *Lenin Peace Prize* in 1957 confirm the resonance of Dolci's actions in the national history of the country.

The exercise of our sociological imagination in the reconstruction developed in these paragraphs allows us to note how Dolci's own action bears witness to his sociological imagination, the quality of mind of a militant who imprints the intersection between biography and history within a society in a cry for help:

> SOS ... here speak the poor people of western Sicily ... they are letting an entire population die out... Sicilians, Italians, men from all over the world, listen... the shacks don't hold up, you can't live in shacks, you can't live on shacks alone. The Italian State has wasted billions in shelters that are obsolete and jumbled

---

separate" in tense and selfish relationships, based on the logic of exchange (Tönnies 1887, It. trans. 1963).

together: but by now the whole area could already be rebuilt, with real houses, roads, schools, hospitals ... SOS... SOS.[11]

9.2 The historical-sociological understanding of a social movement based on non-violent extra-institutional tactics

As mentioned in the *Introduction*, the public issue of the social structure of the Valle del Belìce represented the substratum of a social conflict that saw civil society clash with Institutions – ordinary people in conflict with those who exercise power – a conflict configured as a social movement, regarding which we will put forward a historical-sociological understanding following the logical four-stage process of the Weberian causal analysis (table 1).

The nature of social movement can already be traced in Dolci's SOS, with which we concluded the previous section, since it represents the emblem of an organized, continuous and collective attempt made by relatively powerless individuals (Croteau and Hoynes 2018) to bring the problem of the Valle del Belìce to the attention of institutional channels, in the corridors of power, through a repertory of protests that saw the use of non-violent extra-institutional tactics. On October 14, 1952, Dolci began the first of numerous fasts on the bed of a child who had starved to death; in January 1956, a thousand people began a public fast against illegal fishing, and in the same year hundreds of unemployed people went on an *sciopero alla rovescia* (reverse strike), pledging to reactivate a public road that could not be used because it had not been looked after.

The march of protest and hope for peace and development in Western Sicily took place in 1967; on 25 March 1970, *Radio Libera Partinico* was broadcast illegally for almost twenty-seven hours; on 28 November 1971, three hundred thousand people demonstrated against Fascism in Rome on the initiative of the *Centro Studi e Iniziative*.

The repertory of protests of the social movement promoted by Dolci was rooted in faith and commitment to the affirmation and recognition of universal principles and rights, with the intent to promote change from the *bottom-up*, thus expressing its strength through the *logic of witness* (Ibidem): the intent was to awaken the conscience of the *poor people* in the direction of social participation in institutions, in opposition to facts or social forces (in the Durkheimian sense) based on oppression and evil.

---

11 Extract from Danilo Dolci's introductory message in the broadcast of Radio Libera Partinico. https://danilodolci.org/archivio/radio--libera/

Table 1 – The study of the social movement of the Belìce Valley through the four phases of the Weberian causal analysis

	Weberian causal analysis phases	Valle del Belìce: social movement
1.	Construction of the characteristics of the historical individuality that one wants to explain.	Civil society vs Institutions Non-violent extra-institutional tactics Bottom-up logic
2.	Analysis of the antecedents of historical individuality.	Accidents and situations: – *questione meridionale* (the cultural and economic divide between north and south of Italy) – the 1968 earthquake The effectiveness of individuals/ decisions: – Dolci's horizontal social mobility – Non-violent militancy
3.	Unreal modification of one of the antecedents.	What would have happened if... Danilo Dolci had not been transferred to Sicily... if he had not previously been a militant in other parts of Italy... if he had not proposed innovative tactics of protest?
4.	Comparison of unreal becoming with real evolution	

This message – the faith to be able to generate change from the bottom of people's culture and mindsets – was transmitted through innovative channels for that historical period; the communicative circularity of reciprocal maieutics, the counter-information through the free broadcasting that was illegal at that time, and the graffiti on ruined buildings ("Peace", "He who keeps silent is an accomplice") testify to civil disobedience and its non-violent action against the established order, also through practices that exposed them to repression.

Reciprocal maieutics is a method of "popular self-analysis" based on communicative reciprocity, and shared creation, which stimulates the involvement and active participation of individuals as resources for structural, lasting, and contextual change. This method was applied in the *Centro Studi e Iniziative* (Centre for Studies and Initiatives), established in 1958 in various Sicilian towns, in the *Centro di Formazione per la pianificazione organica* (Training Center for Organic Planning) at Borgo di Trappeto, and in the *Centro Educativo* (Training Center) of Mirto founded in 1975. The *Centro Studi e Iniziative* (Centre for Studies and Initiatives) provided for the involvement of different professional figures (agricultural technicians, social workers, educators and young volunteers) in the search for solutions, tailored to the characteristics of those territories, to face the economic and social situation of western Sicily. The *Centro Studi e Iniziative* in Partinico was responsible for the proposal for a new regional plan called *Città-territorio* (City-territory) for the post-earthquake reconstruction of the Valle del Belìce, based on a bottom-up logic expressed by the involvement of the population and local administrations and the careful evaluation of the physical and social configuration of the context (Cannarozzo 1996).

In the same way, Dolci introduced circular information in mass communication, establishing a free radio – *Radio Libera Partinico*, or *Radio dei poveri cristi* (the Radio of the poor people) – for counter-information from the bottom-up, in opposition to the dominant information of the regime and the monopoly of the State in the control of information and cultural and social reproduction, against «that mass communication that becomes a power structure [becoming] violence, because it imposes the sacrifice of the word to those excluded»[12]: the radio that gives voice to the testimonies of children, women and men who were earthquake victims and abandoned in shacks.

The SOS on the *poor people*, the *Città-territorio* (City-territory) project for post-earthquake reconstruction, il *Centro Studi e Iniziative* (the Centre for Studies and Initiatives) reveal the *framing*[13] of this social movement, since they convey the injustices of the conditions in the Valle del Belìce, propose a viable alternative for post-earthquake reconstruction and remedies for *la questione meridionale* – the cultural and economic divide between north and south, and put pressure on those in the political sphere regarding this public problem.

We have tracked down the breeding ground of this social movement to some general circumstances (accidents), such as la *questione meridionale* (i.e. the cultural and economic divide between north and south of Italy),

---

12 Novara D. (ed.) *Danilo Dolci: omaggio a un maestro*. Available on: https://centrostudialeph.it/archivio/dolci/web_site/dda/omaggio.html
13 «[...] the act of interpreting and assigning meaning to events and conditions in order to shape the message of a movement and the collective identity that develops among its [social movement] members» (Croteau and Hoynes 2018: 440).

exacerbated by the 1968 earthquake. However, we link the characterization of the social movement to the horizontal social mobility of Dolci from the north to the south of Italy and his non-violent militancy.

Dolci was born in 1924 in Sesana, but moved – first for family reasons, then for his social activism – to different parts of Italy, before dying in Sicily in 1997. His training includes university courses in architecture, although he never got a degree.

Dolci soon developed an aversion towards the established order – Fascism – manifested also through his refusal to wear the republican uniform in 1943, and then went so far as to define his action after World War II as a "continuation of the Resistance, without shooting", so much so that the press called him "the Italian Gandhi".

His civic commitment began in Nomadelfia, a city that hosted a community for war-displaced children; he continued by participating in 1951 in the foundation of a new community headquarters in Batignano, near Grosseto in Tuscany. In 1952, he moved to Sicily, where his civic commitment developed[14].

The conditions of misery of that land and the social impact of the 1968 earthquake represent in our analysis that degree of indeterminacy that leads us to ask ourselves: would another man, instead of Dolci, have acted in the same way in the face of the public issue of the Valle del Belìce? If Dolci had not moved to Sicily, would this issue have given rise to a social movement for the claim of the cherished values of those communities? If Dolci had not been a non-violent militant before moving to Sicily, would the social conflict of that land have taken the form of a social movement? If Dolci had not used innovative, even "illegal" protest instruments, such as broadcasting from a free radio station not yet regulated by the State, would this social movement have had the same national resonance? Without Dolci's non-violent civic commitment, would the repressed men, women and children of the Valle del Belìce have claimed their cherished values through fasting, the 'reverse strike', public demonstrations, and free broadcasting?

In the Weberian sense, our answer is *probably not*, since we have no historiographical elements that allow us to suppose and/or affirm that this social movement has been favored by other men, by other decisions, or by other factors. Therefore, we think that Dolci's horizontal social mobility and his non-violent militancy, which began years previously and in other

---

14 His biographical information is dealt with by G. Barone, *La forza della nonviolenza: Bibliografia e profilo biografico di Danilo Dolci* (Napoli: Dante & Descartes, 2004).

places, very likely made the people and institutions aware of the social conflict that was already simmering in those lands – where there was an institutional void in response to the miserable conditions of the South, influenced by the social structure itself and the presence of the Mafia power, which made oppression and submission a social force – characterizing it as a social movement based on non-violent extra-institutional tactics that saw civil society and institutions opposing each other.

Concluding considerations: the impact of the social movement analyzed regarding the birth of free broadcasting

Following the methodological indications of Weberian causal analysis, we have described the essential characteristics of the social movement analyzed (based on the typing of social movements described in the literature) and traced its antecedents to Dolci's horizontal social mobility and non-violent struggle.

Moreover, with this reconstruction work of revealing social and historical material, we have highlighted that, if on the one hand, the understanding of a social movement required the exercise of our sociological imagination, on the other hand, the same understanding revealed that this quality of mind was at the basis of the social movement analyzed: paraphrasing Mills' words, did Danilo Dolci not use sociological imagination, *leading the personal hardship of individuals back to the objective unrest of society, and turning public indifference into interest in public problems*?

The reconstruction work of unveiling made by Severino in the *Intervista impossibile a Danilo Dolci* – Impossible interview with Danilo Dolci – led us to a condition of *retrotopia*, since, through our reflection on Dolci's civic commitment in Sicily we shifted and reassessed the past among the credits, «rightly or wrongly, as a space in which hopes are not yet discredited» (Bauman 2017). The reference to Dolci's civic commitment and the interweaving of biographies, history, and social structure of the communities of the Valle del Belice represent the scenario of a social conflict, but at the same time an example of the promotion of social change.

Unfortunately, we cannot comment either on the level (local and national) or on the degree (circumscribed or radical transformations) of change obtained by the social movement analyzed: the cultural and economic divide between north and south is still ongoing, albeit with differently defined features; most of the social problems that existed then still exist today; the communities of the Valle del Belìce still live in the *non-places* built in the post-earthquake period.

However, we know that *Radio Sicilia Libera* (a.k.a. *Radio Libera Partinico*, a.k.a. *Radio dei poveri cristi* – Radio of the poor people) is unequivocally the forerunner of free radios in Italy, the «first experience of counter-information» (Vitale 2008: 4), which leads us to think that, together with other subsequent Italian attempts, it has favored the birth of free broadcasting in our country (also attributing the adjective "free", subsequently used in institutional law). In fact, this was sanctioned by RAI reform law no. 103 of 24 April 1975, following sentences no. 225 and no. 226 1974 of the Constitutional Court: from that moment, under government authorization, private individuals were granted the right to carry out sound and television broadcasting activities by means of a one-room cable in the local area (Carloni 1995). Thus,

> The highest example of the opening of these new spaces in the airwaves will be given by Radio Aut, the radio of Peppino Impastato, who knew Danilo, sensed his charisma and collected some of his intuitions, primarily that of appropriating the tools usually controlled by those who hold power and use them to preserve it (Vitale 2008: 6).

The concatenation of factors, recalled both in the interconnection between biography and history in a given society, and in the historical-sociological understanding of the social movement we have analyzed, as well as in its contribution to the birth of free broadcasting highlighted in this conclusion, has allowed us to emphasize the reciprocity of dialogue between sociology and history, which, like a "lunar lemon" – an element of identification used by Dolci to metaphorically illustrate his constant civic commitment (Dolci 1970) – never rests.

Bibliographical references

Aron, R., *Main currents in sociological thought: Durkheim, Pareto, Weber*, 2 vols. (London: Penguin Books Ltd., 1967).
Augé, M., *Non-lieux. Introduction à une anthropologie de la surmodernité* (Paris: Le Seuil, 1992); trans. by D. Rolland, *Non luoghi: Introduzione a un'antropologia della surmodernità* (Milano: Elèuthera, 2005).
Bagnasco, A., Barbagli, M., and A. Cavalli, *Corso di sociologia*, 3rd edn (Bologna: Il Mulino, 2012).

Barone, G., *La forza della nonviolenza: Bibliografia e profilo biografico di Danilo Dolci* (Napoli: Dante e Descartes, 2004).
Bauman, Z., *Retrotopia* (Bari-Roma: Laterza, 2017).
Cannarozzo, T., 'La ricostruzione del Belìce: il difficile dialogo tra luogo e progetto', *Archivio di studi urbani e regionali*, 55 (1996).
Carloni, R., 'La formazione del sistema radiotelevisivo misto: le indicazioni della Corte costituzionale e le risposte del Legislatore', *Bollettino di informazioni costituzionali e parlamentari*, n. 1-3, (1995), 225-240.
Cascavilla, M., 'La sociologia del diritto penale di Émile Durkheim', *Studi di sociologia*, n. 3 (2018), 273-290.
Cavalli, A., 'Il rapporto tra conoscenza storica e sociologia in Max Weber', *Il politico*, 45 (4) (1980), 571-590.
Croteau, D., and W. Hoynes, *Experience sociology – culture, structure, power*, 3rd edn (New York: McGraw Hill Education, 2018); ed. by F. Antonelli and E. Rossi, *Sociologia generale: Temi, concetti, strumenti*, 2nd edn (Italy: McGraw Hill Education, 2018).
De Nardis, P., 'Prefazione', in S. Severino, *Intervista impossibile a Danilo Dolci: Saggio sulle funzioni della radio per lo sviluppo dei fatti sociali* (Roma: Aracne editrice, 2015), pp. 11-18.
Dolci, D., 'Anche le piante dopo scaricate', in D. Dolci, *Il limone lunare: Poema per la radio dei poveri cristi* (Bari: Laterza, 1970), pp. 190-191.
Durkheim, É., *De la division du travail social* (Paris: Alcan, 1893); It. trans. *La divisione del lavoro sociale* (Milano: Comunità, 1962).
Giddens, A., *The consequences of modernity* (Cambridge: Polity, 1990); It. trans. *Le conseguenze della modernità* (Bologna: Il Mulino, 1994).
Gubitosa, C., *Radio Libera Partinico, storia di una voce scomoda* (libretto autogestito, 16 May 2005).
Mills, C.W., *The sociological imagination* (New York: Oxford University Press, 1959).
Vitale, S., 'S.O.S. ... S.O.S. ... qui si sta morendo', in G. Orlando, and S. Vitale (eds), *La radio dei poveri cristi: Il progetto, la realizzazione, i testi della prima radio libera in Italia* (Marsala (TP)-Palermo: Navarra Editore, 2008).
Pescosolido, G., 'La questione meridionale', in *Dizionario di storia* (2010), http://www.treccani.it/enciclopedia/la-questione-meridionale_%28Dizionario-di-Storia%29/ [accessed 15 July 2020].

Severino, S., *Intervista impossibile a Danilo Dolci: Saggio sulle funzioni della radio per lo sviluppo dei fatti sociali* (Roma: Aracne editrice, 2015).

Severino, S., 'Ipostatizzazione dell'identità meridionale', in N. Arrigo, A. Gaudio and S. Severino (eds), *Ripensare il Sud: Contraddizioni e possibilità di un nuovo umanesimo* (Foligno: Rivista di Studi Italiani, 2016), pp. 19-35.

Tilly, C., *Social movements 1768-2004* (Boulder (CO): Paradigm, 2004).

Tönnies, F., *Gemeinschaft and Gesellschaft* (Leipzig: Reisland, 1887); It. trans. *Comunità e società*, (Milano: Comunità, 1963).

Topitsch, E., *A che serve l'ideologia?* (Bari: Laterza, 1975).

Weber, M., *Gesammelte Aufsätze zur Wissenschaftslehre* (Tübingen: Mohr, 1922); trans. by P. Rossi, *Il metodo delle scienze storico-sociali* (Torino: Einaudi, 1958).

CINZIA RECCA

10.
NAPLES AND THE REVOLUTIONS OF 1799:
A CLARIFICATION OF THE FACTS BY TRUTHFUL ACCOUNTS
OF SOME EYEWITNESSES

In every period of history, certain turning points are particularly significant and rich in changes. During early modern European history, all roads led to a revolution, some ages more than others. The initial idea of the social revolt and then the actual event predominated especially during the 17th and 18th centuries[1]. From the end of the 15th century to the beginning of the 17th, Europe was in upheaval. However, it must be said that the "crisis of the 17th century" did not affect all of Europe, but old Mediterranean European countries – such as Spain – which were still linked to chivalric and feudal models[2]. Instead, during the 17th century, the more "bourgeois" nations that faced the Atlantic, such as England and Holland, experienced a phase of great economic expansion. The world changed after the geographical discoveries: when the Mediterranean was the centre of Europe, the nations on the Mediterranean sea such as the Sicily of Frederick II (ruled by the Normans and Aragons) were rich; however, when the centre of Europe moved to the Atlantic – following the geographical discoveries – Mediterranean Europe entered a crisis from which it never recovered[3].

---

1   See Nicolas, J. *La Rébellion française. Mouvements populaires et conscience sociale 1661-1789* (Paris: Seuil, 2002). In the decade 1764-1774, he analyzes 367 episodes of turmoil originating from the *Guerre du blé*.
2   There is an impressive bibliography on the subject. For the coordinates of the historiographic debate of the twentieth century, see the important contribution of Francesco Benigno, Benigno F. *Specchio della Rivoluzione: conflitto e identità politica nell'Europa moderna* (Roma: Donzelli, 1999). Very recently Bailey Stone published an interesting and innovative study offering a balanced and nuanced examination of both structuralist and postmodernist theories focusing on new ways of understanding radical change in the European polities that created the concept of modern revolution and its dramatic realities: Stone, B. *Rethinking revolutionary change in Europe: A Neostructuralist Approach* (Lanham- Boulder-NewYork London: Rowman & Lietfield, 2020).
3   Romano, R. *Opposte congiunture. La crisi del Seicento in America e in Europa* (Venezia: Marsilio, 1992); Parker, G. *Global crisis: War, Climate Change and Catastrophe in the Seventeenth Century* (New Haven:Yale: University Press,

This difference between the North and South of Europe also influenced population growth: whereas in Mediterranean Europe the population decreased, in Holland and England the population increased, causing famine and several popular insurrections between1647-1648. Albeit in the Mediterranean region, the most famous revolt was that of Naples in 1647[4].

The number of changes that took place in Europe around the late 1750s and early 1760s is quite remarkable. This was an age of rapid social and political changes, the most significant of which are the already mentioned increase in the population, the agrarian and industrial revolutions, the growth of a wealthy bourgeoisie, the Enlightenment, despotism, the challenge to the monarchy and aristocracy, and popular revolt (Benigno 1999). It was a century that ended with the industrial revolution and the great political revolutions in France and Naples.

Before 1799, Naples was the third most important city in Europe after London and Paris, for its demographics and the fact that it was an enlightened cultural and artistic landmark. Indeed, Naples was considered a land of immigration: the splendors of the big city, the port, the arsenal, the construction sites, the Royal Court and the courts of the numerous noble families, and the probability of survival in the event of famine favored the explosive growth of the population[5]. In the early 1790s, most of the municipalities of the Kingdom of Naples were state-owned, mostly subject to royal jurisdiction, and there were also feudal and ecclesiastical municipalities (Ajello 1961). There was a strict social hierarchy comprising several social classes from noblemen to common people. The noble class together with the military represented the most visible part of the population, who participated in court life and did not need to work to

---

2013); A detailed bibliographical reconstruction is provided by Abulafia, D. *The great Sea: A Human History of the Mediterranean* (London: Penguin, 2015). As regards Italy, see Malanima, P. *La fine del primato. Crisi e riconversione nell'Italia del Seicento* (Milano: Mondadori, 1998).

4   On this revolt and its several interpretations, see among others: Villari, R. *Un sogno di libertà. Napoli nel declino di un impero. 1585-1648* (Milano: Mondadori, 2012); Hugon, A. *Naples insurgée 1647-1648: De l'événement à la mémoire* (Rennes: Presses Universitaire de Rennes, 2011*)*; Di Franco, S. "Le rivolte del Regno di Napoli 1647-48 nei manoscritti napoletani", *Archivio Storico per le Province Napoletane*, CXXV, 2007, pp. 1-132; Di Franco S. and Musi A. eds., *Mondo antico in rivolta (Napoli 1647-1648)* (Roma-Bari: Lacaita, 2006).

5   The population of the city was around 550,000. See: Di Battista, F. *Il Mezzogiorno alla fine del '700* (Roma Bari: Laterza, 1992); Mafrici, M. (ed), *Il Mezzogiorno settecentesco attraverso i catasti onciari*, 2 vols (Napoli: Edizioni Scientifiche Italiane,1986).

live (Battaglini 1996:115-116). The situation of the middle classes, made up of the professional and propertied class, was still precarious. Even if they were growing in dimension and confidence, thanks to their cultural ambition cultivated in the university, the monopoly of elderly bureaucrats devitalized them (North 2018:17-29).

10.1 *The Neapolitan revolution and the controversial role of the populace*

On January 22, 1799, French troops forced their way into Naples and in doing so they confirmed the authority of a Neapolitan Republic which was proclaimed one and indivisible the day before by a group of patriots who had taken control of Castle Sant'Elmo (the fortress on the hill immediately above the centre of the city). This was the last of the European revolutions and has been regarded as the offspring of the great French revolution. Like its predecessors in Northern Italy and elsewhere, it strongly depended on French military intervention. In fact, during the weeks prior to January 22, the patriots were unable to control the city, and popular violence and disorder reigned supreme (Pietrabondio 1884).

As Michelle Vovelle points out, the revolution in Naples took the form of dramatic clashes and involved large categories of the population. However, the Neapolitan revolution was an ephemeral blaze of five months; it came after the French revolution, and it cannot be said to have been incited by the troops of the coalition because it was already taking place when they arrived (Rao 2011:855-860). Therefore, the French intervention only served to ensure the success and proclamation of a Republic, despite the internal and external pressure of a counter-revolution. Consequently, the Neapolitan Republic of 1799 was not simply a figurehead of French military occupation, but an effect of a violent revolution caused not by the French but by the actions of the immature and feeble rulership of King Ferdinand IV of Bourbon and his wife, and at the same time by the passionate desire of the enlightened intellectual class to be released from a tyrannical dynasty.

The paradox pointed out by several historians consists in the urban setting (Vovelle 1999:358-369). What was the revolution of the masses without the masses? We could affirm that the Naples revolt was characterized by the weakness of the revolutionary movement. One of the leading lights of the Neapolitan Enlightenment was without doubt Vincenzo Cuoco, who believed that the revolution failed because it was a passive revolution, in which a small group of republicans had tried to force French ideas on a people who were mostly unaware of their meaning and unprepared to receive them. The republican infatuation with the French Revolution with its grand ideas and conspiratorial clubs meant that they neglected to involve the populace in their programme and thus failed to make the revolution active (Cuoco 1998: XVI-XVII). Therefore, the enlightened *elite* was not sufficient to defeat the enlightened despotism of the Bourbons that came back triumphing and claiming a bitter vengeance. The short experience ended tragically in the reconquest by the Sanfedists, the Bourbon royal troops led by Cardinal Ruffo, and culminated in a general massacre[6]. After the tragedy of 1799, Naples became one of the capitals of counter-revolution.

No writer has written about that period with the impartiality that the dignity of history requires. There are many accounts and questionable narratives of historians, but certainly the chronicles of the actors of that period remain valuable testimonials. There are some literary texts about the historical characters and heroic dreamers of the Neapolitan revolution, who were united by the illusory attempt to seek good for a deaf and indifferent populace (Croce 1912).

'A Historical Essay about the Naples Revolution of 1799' is one of Cuoco's first papers (Cuoco 1998). This first large work with a historical patriotic character is one of the most important 19th century accounts of political revolution and shows great ability for self-criticism:

> But a physical catastrophe is most often more accurately observed and more truthfully described than a political catastrophe. I narrate the events of my homeland; I tell of events that I myself have seen, and of which I was myself one day not the least part: I write for my fellow citizens that I must not, cannot, and do not want to deceive (Cuoco 1998:6)

Another remarkable work is Francesco Lomonaco's *Report to citizen Carnot*, an essay reporting the repression started by Ferdinand IV and inviting patriots to remove the King from the political scene (Lomonaco 1999). The essay is rich in observations and unscrupulous criticism against the Bourbon Crown, and it could be summed up in this statement: «The hypocritical tyranny has succeeded in spreading the blight of discord and

---

6 On General Fabrizio Ruffo, see: Casaburi, M. *Fabrizio Ruffo: l'uomo, il cardinale, il condottiero, il politico* (Roma: Rubettino, 2003); Ruffo, G. and De Maio, D. *Il cardinale Fabrizio Ruffo tra psicologia e storia: l'uomo, il politico, il sanfedista* (Roma: Rubettino, 1999); Sacchinelli, D. *Memorie storiche sulla vita del Cardinale Ruffo* (Napoli: Cataneo, 1836).

civil war and arming citizens against each other» (Ibid 28). Lomonaco took part in the Neapolitan Republic and succeeded in escaping from the Bourbon repression before seeking refuge in France (Rao 1992).

However, these eminent political intellectuals have largely been analysed in historiography so we prefer to pay attention to the testimonies of less famous figures than the ones mentioned, highlighting the accounts of the Neapolitan revolutionary period by the cleric Drusco Pietrabondio, the archpriest Francesco Apa, Domenico Petromasi (a lieutenant colonel of the royal army), and Giuseppe De Lorenzo, a national Royal guard. These men recorded their experiences in two distinctive periods of 1799 (Sanguinetti 2011): the Neapolitan revolution and the counter-revolution marked by the reconquest of the Crown.

Drusco Pietrabondio lived through and recorded the terrible period of the Neapolitan revolution (from December 21, 1798 until January 23, 1799); he depicted an anarchical situation which had reached a state of complete violence in January 1799. The entire city seemed in revolt against the Jacobins and, according to Drusco, the populace defined *plebe* – plebeians – were in any case naturally quite as "mad" as those who held the power (Pietrabondio 1884: 23).

Francesco Apa recorded his experience from March 17 until June 13, 1799. He was anti-Jacobin, faithful to his King, and a supporter of Fabrizio Ruffo's troops (Apa 1800:18). His memoirs, written out of duty to his Sovereign, are particularly important since they were written in the neo-republican period and provide us with the record of an archpriest nominated general inspector of food and fodder. His work shows his excellent knowledge of history and linguistics, and some news stories that he describes correspond with the dates. His perception seems to be direct and not mediated; the society is metaphorically divided into lambs, wolves and the lions: «Separating the lambs from the wolves would purge the kingdom of the seeds, preventing contagious disease from planting its roots» (Apa 1800: 8). The lions were the ones who were in favor of Cardinal Ruffo bringing back the monarchy to Naples and so the revolution was almost denied: the people democratized themselves in the province for fear and terror (Ibid 37). The role of Apa in this frantic period, as he wrote, was intended to arm many populations of his diocese, and he took care to prevent the arrival not only of external enemies, but also of the bad people of the Kingdom.

> If only he had seen the people without his humble pity, many places democratized themselves only out of terror, and fear, which had the influence of threatening voices [...] He had had many people in his diocese armed en

masse ... and took care to sever ties not only with the enemies coming from outside but also the evildoers of the kingdom (Ibid 28).

He guided a troop describing it as an unruly crowd, stating the way to guide it with a particular metaphor: « They should be treated like the pupil for my eyes» (Ibid 35).

Another ardent royalist and actor during the counter-revolution was Domenico Petromasi, a war commissioner, lieutenant colonel, and a very faithful follower of Fabrizio Ruffo. He also wrote a chronicle of the reconquest of the Bourbons, and his version of the events of Naples in 1799 is distorted by his position which filters the perception of events[7]. In Petromasi's opinion, the people had been deceived because they were not truly free because they had passed into French hands:

> But if a people said they were free if they were not subjected to the power of others, how could the Neapolitan people, who were declared subject to the dominion of France by the words of the proclamation itself, believe themselves free?

There were those who explained the status of free beings referring to the past; however, whatever the interpretation, it caused unexpected and harmful effects[8]. In fact, even if Petromasi and other Sanfedists who wrote their memoirs truly lived through the counter-revolution, we cannot consider them to be reliable because they wrote their works to show their efforts and obtain recognition from the Bourbon King[9].

A memoir written in a different tone, less partial and so more authentic, is that of another army man, Giuseppe De Lorenzo, a national guard of the Neapolitan Republic during 1799[10]. He was not a political intellectual or an

---

7   On some biographical statements of Petromasi, see Petromasi, D., *Alla riconquista del Regno di Napoli. La marcia del cardinale Ruffo dalle Calabrie a Napoli*, intr by Silvio Vitale (Napoli: Il giglio, 1994).
8   Here Petromasi is implicitly referring to the Neapolitan republican circles. Ibid, 124.
9   This is evident in the first pages of Petromasi's work, in which he addresses the King directly with adulation: Petromasi, D. *Storia della spedizione dell'Eminentissimo Cardinale D. Fabrizio Ruffo allora vicario per S.M. nel Regno di Napoli e degli avvenimenti fatti d'armi accaduti nel riacquisto del medesimo compilata da D. Domenico Petromasi*(Napoli: Vincenzo Manfredi, 1801), IV-V.
10 On biographical statements of Giuseppe De Lorenzo, see Massafra, A. *Patrioti e insorgenti, in provincia: il 1799 in terra di Bari e Basilicata: atti del Convegno di Altamura-Matera: 14-16 ottobre 1999* (Bari: Edipuglia, 2002). 51- 55; De Lorenzo, G. *Nel furore della reazione del 1799*. Prefazione e note di B. Croce

activist, but at the age of twenty-one he decided to enlist as a civic guardian because he was driven by a strong personal desire to work for himself and for his family, as he himself wrote «I propose to give you a short account of the most remarkable events... ...of those that have happened to me in the change of government three times in my unhappy country» (De Lorenzo 1999:247).

The most interesting aspect of this memoir lies in the description of the counter revolution, a violent reaction of the plebs and the military Bourbon troops supported by Sanfedista masses, and the horrifying torments suffered by those who were persecuted and taken as Jacobins, as he wrote:

> But when the joy turned into horror (...) we met a crowd of brigands and armed thugs, all intent on ruthless looting, and who were judged by the people to belong to the houses of the Jacobin patriots (...) who were led under arrest by thugs, after having been intimately stripped and wounded (...), women, mothers, young girls and spinsters were taken by the people in the naked procession, because they belonged to the family of some Jacobin (Ibid, 262)

De Lorenzo describes the populace in revolt and ruthless against the "Jacobin enemy", demanding the return of their King. The people are described as "powerful" and "armed", ready to vindicate the Bourbon monarchy. So, from his detailed description of the acts of vindication and cruelty of the Neapolitans (Ibid 264), we have an almost contradictory image of the people of Naples described by Cuoco. As two sides of the same coin, passivity and spirit of revolt are embodied in the low-class Neapolitan people. During the Neapolitan Republic, the people did not want to listen to the government because they did not understand the republican thought, and even if there were several attempts to instruct the people, they refused them, preferring obedience to their beloved King.

10.2 *Misunderstood chronicles of female revolutionaries*

In addition to all the historical documents that have been handed down to us in which the exploits of male characters (heroes and martyrs) are always emphasised, we would like to highlight those of three women who lived through the revolution and examine the ways in which these women

---

(Napoli: Colonnese, 1998); De Lorenzo, G. *Memorie*, edited by Paola Russo (Napoli: Vivarium,1999).

perceived the images of the revolution through their thoughts and memoirs written at that time to transmit political and cultural messages.

Eleonora Pimentel Fonseca and Luisa Sanfelice are two female figures whose courage, temperament, and strong will were equal to those of their male counterparts. They were both Jacobins and revolution martyrs, and as such represent the other side of the Neapolitan revolution in 1799. Eleonora Pimentel Fonseca was the director of the newspaper *Monitore Napoletano*. Her verdict of guilty was requested by Maria Carolina of Austria, Ferdinand IV's wife because of her papers against the Monarchy and lese Majesty. Luisa Sanfelice was described by several historians as one of the Neapolitan revolution heroes who gave up her life for the sake of freedom.

Eleonora de Fonseca Pimentel, poetess, scholar of jurisprudence and natural and mathematical sciences, was initially an enthusiastic supporter of Ferdinand IV and the Bourbon's enlightened political reforms[11]. After the French Revolution and the radical change of the Neapolitan government policy, which became much more illiberal and repressive, she became an active proponent of the 1799 revolution and leading figure of the "Neapolitan Republic". She also was a free, courageous journalist and founded a republican newspaper *Monitore Napoletano*, which was

---

11 On Eleonora De Fonseca Pimentel (Rome, 13.01.1752- Naples, 20.8.1799) see: Orefice, A. *Eleonora Pimentel Fonseca. Eroina della repubblica napoletana del 1799* (Roma: Salerno Editore, 2019); Fonseca Pimentel, E. *From the Arcadia to the Revolution. The Neapolitan Monitor and other Writings* ed. and trans. Jones V.R. (Toronto: Iter Press, 2019); Pellizzari, R.M., "Eleonora de Fonseca Pimentel: morire per la rivoluzione", *Storia delle Donne*, no.4 (2008), 103-121; D'Episcopo, F. *Eleonora Fonseca Pimentel tra mito e storia* (Napoli: Edizioni Scientifiche Italiana, 2008) ; Rao, A.M. "Eleonora De Fonseca Pimentel, le monitore napoletano et le problème de la partecipation politique"*, Annales historiques de la Révolution Française*, 344 (2006): 179-191; Santos, M.T. "Leonor da Fonseca Pimentel. A Portuguesa de Nápoles (1752-1799)" *Actas do colóquio realizado no bicentenário da morte de Leonor da Fonseca Pimentel* (Lisboa: Sara Marques Pereira, 2001); Forgione, M. *Pimentel Fonseca* (Roma: Newton & Compton, 1999); Urgnani, E. *La Vicenda Letteraria e Politica di Eleonora de Fonseca Pimentel* (Napoli: La Città del Sole, 1998); Battaglini, M. *Eleonora Fonseca Pimentel: il fascino di una donna impegnata tra letteratura e rivoluzione* (Napoli: Procaccini, 1997); Macciocchi, M. A: *Cara Eleonora. Passione e morte della Fonseca-Pimentel nella Rivoluzione napoletana* (Milano: Rizzoli, 1993); Buttafuoco, A.R. "Eleonora Fonseca Pimentel: una donna della rivoluzione", *Donna e ricerca storica*, n. 3 (1977): 51-72; Schiattarella, F. *Marchesa giacobina. Eleonora Fonseca Pimentel* (Napoli: Schettini, 1973); de Fonseca Pimentel, E. *Il Monitore Napoletano del 1799. Articoli politici seguiti da scritti vari della stessa autrice* ed. Benedetto Croce (Bari: Laterza, 1943).

committed to changing "Neapolitan plebs" into "Civil people". At the fall of the Republic, she was fearless in meeting death by hanging.

The image of Eleonora Fonseca Pimentel has been depicted through the cultural model of the republican martyr according to the characteristics of ancient Roman virtues: those very values would then be entered in almost all the nineteenth-century biographies of Pimentel. The chroniclers present her on the gallows platform as a woman without a moment of weakness: the same fearless attitudes in the face of death in mythography reserved for the male exponents of the Neapolitan events of 1799. In her writings we can see clear messages in which she expressed her desire to build a more honest and free society through the education of the people. However, we should first examine the most significant aspects of her biography, then return to the symbolic value of her role as revolutionary.

Eleonora de Fonseca Pimentel descended from a noble family. Since her adolescence, she showed various talents and a passion for studying; she loved not only classical literature and poetry, but also scientific studies. Her first teacher and mentor was her uncle, an abbot who encouraged her to attend the most prominent literary salons. At the age of sixteen she recited verses in the literary salons. Her education proceeded on a precise path that took her out of the traditional mold of her gender towards certain social models not so common to the female experience but which, however, had begun to open a window to women on the world of knowledge and public matters.

In 1768, she entered two of the most famous academies, Arcadia and that of the Philaletheans. Ideas regarding drastic social change fermented in the setting of the academies and salons; these tended towards the implementation of a human and social emancipation process in the framework of a liberal and enlightened monarchy, which would have been that of Ferdinand IV and Maria Carolina. Eleonora joined the climate of confidence in the Bourbon Reforms with enthusiasm. The education path of the Arcadian poetess is significant: as an intellectual, she proceeded as any other philosopher and man of letters of her time, who was impressed in Naples during the "heroic period" of the Bourbon dynasty, and who later subscribed to revolutionary ideals beyond the Alps.

During the 1780s there was a peaceful cohabitation between intellectuals and the Bourbon rulers: Eleonora continued to write verses of sincere admiration for government policy, such as those in 1789 dedicated to the creation of the San Leucio colony and a royal factory inspired by egalitarian principles, which caused great excitement among the Neapolitan reformers. The domino effect of the French Revolution had even brought about the overthrow of King Ferdinand IV and Maria Carolina; after twenty years of an ostensibly reformist government, Maria Carolina had turned her back on the classes considered to be the vehicle of the revolution. Consequently, she broke ties with the same reformist forces that had sustained and supported her in her attempts to modernize the state structure of the Kingdom. After the French Revolution, there was a clear fracture between the throne and the most intellectually and politically advanced classes, now considered by the sovereign and her court as enemies. During the early 1790s, the intellectual reformist class was undergoing a process of accession to the principles and revolutionary action. So, at that time the Neapolitan sovereign espoused a series of restrictive, policing measures: abandoning the reform program and switching to the edicts banning the Freemasons, to the agreements with the Holy See, and to defensive and offensive alliance with the sole purpose of preparing for the war against France.

The years 1794-795 must have been crucial for Eleonora for the intellectual and political crisis that culminated in her emotional reaction to the death of the young authors of the Jacobin conspiracy. As Benedetto Croce pointed out, she belonged to those idealists who saw the government moving away from the program that until then had been followed and cheered, and at the same time she discovered the plan of the revolutionaries in France. Therefore, moving away from the old hopes and the old methods of collaboration with the government, the Neapolitan intellectuals acquired faith in the new methods, in a process that transformed them from royalists to revolutionaries. For Eleonora, a period began when she was suspected of having relationships among the patriots, a period that culminated with the suspension of royal aid on her behalf in 1797, and with her arrest on October 5, 1798. She was released from prison in early January 1799, and it is almost certain that she immediately attended meetings of the Committee of the patriots who advocated the establishment of a democratic republic. Eleonora took possession of Castle Sant' Elmo with the other patriots, becoming a proponent in the events of the Neapolitan Republic and opening the way for the French in Naples.

After experiencing prison and seeing fighting and people dying around her, the praised poetess of the past now entered into a personal struggle, and exposed herself by professing the cause of freedom and the defeat of those now identified in the image of the tyrants, the Bourbons of Naples[12].

---

12 Some pages of the journal Neapolitan Monitor could be seen as a real invective against the government of the Bourbon Crown. To this regard, see the first editions of *Monitore Napoletano*, specifically, numbers 4,6,7, and 11.

In 35 numbers of the republican journal *Monitore Napoletano* she showed her strong personality, expressing her independence in the face of any pressure from the Monarchy. Her main intention was to protect the interests of the country and above all the people. She wanted to protect them against the royal tyranny through the transmission of the knowledge of the new concept of society, and the first purpose of the newborn Republic was to instruct the people about the new concept of society, as she reported in the pages of the *Monitore*:

> Our government calls for the education of the people [...] A people who pass in a stroke from slavery to freedom cannot be completely reborn to such a happy state if uniform instructions of hard morals and true patriotism do not form the spirit, the public custom, and the true support of good laws equally in all Individuals. They should be instructed in the shortest time possible with a Catechism of morals for the intelligence of the people (*Monitore Napoletano* n. 31, 25 May 1799, Fonseca Pimentel 1943)

She lived for the *Monitore* whose editorial staff was firmly established in her own house, so in the months of the republic her life was dedicated to the entire direction of the newspaper[13]. She wrote most of the articles herself, publishing information, as well as collecting it directly, participating in the meetings of the government as well as in the events and ceremonies of the Republic. Disregarding all the alarming rumors of a counter revolution that came from the Neapolitan suburbs such as Casoria and Torre Annunziata, Eleonora continued to write with the intent to always leave the public confident in a republican victory. Up to the end she incited the people to take up arms for freedom and independence. The first seed of Italian unity was sown in Naples with its revolution.

Eleonora exhorted the people to fight, to abandon all their possessions 'houses and substances', because everyone must have one faith, "all must believe in salvation, and only through freedom would they be saved". Her dream ended with the end of the Republic, and with it her life ended too. On August 17, she was sentenced to death by hanging in front of the plebians – those same "plebs" for whom she had had the great project of education: her Jacobin political dream of educating the populace to freedom and a common sense of civil and high egalitarianism.

Besides the intellectual and political activist Eleonora Fonseca Pimentel, we have the historical evidence of the Neapolitan Revolution

---

[13] On the presentation and analysis of Neapolitan Monitor, see the recent study edited and translated by Jones, V. R. *Fonseca Pimentel* (2019).

of 1799 of Luisa Sanfelice, another woman who distinguished herself for her revolutionary ideals, whose moniker "the lost girl" was later misrepresented by historians. Nevertheless, her personality and her tragic end made her a star of the 18th century: beautiful and passionate, the victim of a wayward husband, Luisa was involved in a counter-revolutionary conspiracy by the Baccher brothers. After the return of the Bourbons in Naples, she was locked up in prison and sentenced to death. After a long ordeal when her death sentence was postponed and then later confirmed, she was beheaded on September 11, 1800.

The causes of her beheading are not related to her revolutionary thoughts and activities, in which she was involved implicitly and unconsciously due to her love affair with an active Republican, but because her husband Andrea Sanfelice, wanted to take revenge for her various marital infidelities. However, as Benedetto Croce stated, the reported accounts are uncertain and still lacking in evidence: Luisa Sanfelice's unfaithfulness and her political choices are merely unproven suppositions, perhaps made to better understand the events that led her to die on the scaffold.

Fascinated by the sophistication of the salons and parties, Sanfelice frequented monarchist and republican circles indifferently. At that time, Sanfelice was the object of desire of the young Gerardo Baccher, a Bourbon leader who was planning to conspire against the Neapolitan Republic of 1799, and the Republican Ferdinando Ferri. After the instauration of the Neapolitan Republic, she enjoyed the glory of the republicans.

The most common theory regarding the conspiracy affirms that Gerardo Baccher informed Luisa of an upcoming "pro-monarchy" revolt and gave her a pass card to be shown in case of danger. However, fearing for Ferri's life, Luisa preferred to give him the passcard. Ferri informed Vincenzo Cuoco of the conspiracy plan who in turn decided to warn the government. Gerardo Baccher and other conspirators were arrested, including his father and two brothers. Thus, Luisa accidentally became a "saviouress of the Republic and mother of the Nation" (Croce 1942).

In the edition of April 13, 1799 of her journal *Monitore Napoletano*, Eleonora Pimentel Fonseca reported the discovery of the "conspiracy of Baccher" and with these words she acclaimed Luisa Sanfelice, praising the noble deed:

> One of our eminent citizens, Luisa Molina Sanfelice, revealed on Friday evening to the Government the conspiracy of a few fools rather than villains, who, trusting in the presence of the English team, or in collaboration with them, intended on Saturday to massacre the Government, the good patriots, and then attempt a counter-revolution. The head of the foolishly iniquitous project was

a German called Baccher, who was employed by the Abbenanti merchant, and who was arrested that same night, and taken to court the following morning, dragging under his arm the Royal flags, which were found near him. There were also some security cards, which were to be handed out, like similar ones that had been given to those who wanted to be saved, and all the rest of the insurgents were destined for slaughter (in their imagination). It appears that these papers bore the arms of Ferdinand and the English lion at the expense of the barons. (*Monitore Napoletano*, n.19, 13 aprile 1799. Fonseca Pimentel 1943).

Luisa's report in the *Monitore* was read by the Bourbon Royal Couple during their stay in Palermo, and as soon as they returned to Naples, they organized their tragic revenge (North 2018:176-197).

Revolution and counter-revolution are embodied in the figures of Pimentel and Sanfelice. Two emblematic female protagonists of Italian history, Pimentel was a political heroine who became a character able to express her ideas and to transmit them in a society where women were not recognized as citizens. Pimentel started to be seen as a rational being, bearer of liberal ideas through the circulation of the *Monitore Napoletano*, while Luisa Sanfelice, a gorgeous woman often described as an amazon and "lost girl", guided by her passion and courage, became an accidental protagonist.

10.3 *Revolutionary accounts by an English novelist*

The Neapolitan revolution and the figure of Fonseca Pimentel captured the interest of an English novelist Helena Maria Williams who, far from Naples, in her *Sketches on the French Republic*, wrote six letters describing the Neapolitan Court of the Bourbon King Ferdinand IV and the revolution of 1799. First of all, I will briefly outline some key aspects of her life-story, to grasp her revolutionary attitude.

Helena Maria Williams was an English poet, novelist and chronicler of the French Revolution; she vividly recounted her experience in France during the Terror. Helen Maria Williams, as a woman and not a politician, took a position and justified her seemingly unpatriotic interest in the Revolution. Williams had to show her readers that the heart-felt responses that had made people appreciate her emotional poems would also enable her to write about the Revolution without sacrificing her femininity (Kennedy 2002:63). When the French Revolution began, Williams travelled to Paris in 1790 with her mother and sister, Cecilia, to participate in the "glorious spectacle" and celebratory atmosphere of the Revolution. She witnessed many of the key celebrations and political transformations in revolutionary France and chronicled those observations publishing eight volumes of her *Letters Written in France* on revolutionary France from 1790 to 1796[14].

Prior to 1790, she had established herself as a Romantic novelist and poet, using her pen to critique such contemporary issues as the suffering caused by war and the slave trade, and also to promote the success of the American Revolution, especially its establishment of democratic government[15]. By 1792, with the September massacres and the execution of Louis in January 1793, Williams was no longer uncritically enthusiastic about the Revolution. Struggling with a bout of illness and the disapproval of some of her friends in England because of her decision to stay in Paris, Williams had to find new rhetorical strategies in order to write about her experiences beyond the scope of the joyful or pathetic scenes recorded in her first two volumes: namely, regicide, political factions, intrigues, and moral failure. She moved towards establishing herself as a serious political commentator, outlining the massacres and the atrocities like a wartime journalist.

Williams was arrested in the fall of 1793, and she recorded with passion and sorrow the degeneration of the French revolution into chaos and murder. She became a testimony of the most atrocious barbarities. By the mid 1790s, Williams had attained an unprecedented position for a female writer of her day by becoming a well-known authority on an international event of immeasurable historical importance.

Williams' account of the Revolution in her first work describes a series of simple events, and she seemed to refuse to include any violence. Williams used nature, as did the revolutionaries themselves, as a means of legitimating the Revolution. While the revolutionaries planted "liberty trees" throughout France as a symbol of the natural outgrowth of their freedom, in her work Williams strives to show the Revolution as an

---

14   A work of a mixed genre, in the form of travel literature and family letters: Williams H. M. *Letters Written in France, in the Summer 1790, to a Friend in England; Containing, Various Anecdotes Relative to the French Revolution; and Memoirs of Mons. and Madame du* F (London: Thomas Cadell, 1790); Ibid, *Letters From France; Containing a Great Variety of Original Information Concerning the Most Important Events that Have Occurred in that Country in the Years 1790, 1791, 1792, and 1793* (Dublin: J. Chambers ,1794) ; Ibid, *Letters containing a sketch of the politics of France from the thirty-first of May 1793 till the twenty eight of July 1794 and of the scenes which have passed in the prisons of Paris, vol. IV* (Dublin: J. Chambers, 1796).

15   Thus, it is not surprising that when in 1790 Williams published her novel *Julia*, she included a poem, "The Bastille," that praised the ideals of the French Revolution.

embodiment, restoration, or redefinition of the natural – Nature affirms the rightness of the Revolutionary ideals: the aristocracy is portrayed as a distorted and twisted tree, while the tree of patriotism is tall and straight. A key aspect of Williams' identification of revolutionary politics changed and was influenced by the idea of a particularly feminine sensibility in the struggle for human rights. Williams aimed to universalize this sensibility across gender boundaries, opposing the Burkean notion of 'manly morals' by establishing the human hearts of both men and women as the "natural terrain of politics". For Williams, the Revolution and its principles lived on despite the Terror, «like vigorous seeds committed to the fertile earth...they will remain alive, and ready to spring up at the first favourable moment» (Williams 1794: 94).

Throughout her forty-five-year long career, she remained an 18th century woman of letters with an unflagging love for liberty. The success of her letters and the criticism they provoked provided her with the unique opportunity to transform her work from indirect to direct social and political commentary and she became a foreign correspondent, interpreting French history in England and around the continent for thirty years. The 1790 volume imagines the entire Revolution as

> a sublime spectacle carrying forth the spirit of [this] Federation, appealing to the noblest of human sentiments, and establishing aesthetic and moral harmony across differences of sex, race, and condition (Williams 1790:14-15).

For Williams, this historical moment created a vision of a new path not only for France but also for all of Europe, where the establishment of human rights could flourish.

In her first fifteen letters, Williams recounts her journeys from Paris to Rouen, visiting the symbolic origins of the revolutionary transformation from tyranny to freedom. This was a journey that encompassed both the public and the private transformation of tyrannical authority in France. For example, Williams re-created for her readers the historic moment – the fall of the Bastille – that the Federation celebrated by a personal visit to the former royal prison[16]. Although the new political order limited the ranks

---

16 The overthrow of the Bastille represented the Enlightenment of the French and the courage of the men and women who had achieved this historic feat and after having a visit of Bastille, Williams wrote: "we may indeed be surprised, that a nation so enlightened as the French, submitted so long to the oppressions of their government; but we must cease to wonder that their indignant spirits at length shook off that yoke". Williams (1794), 74.

of future deputies through stringent economic requirements, the Assembly opened its tribunes and its podium to all. Williams marvelled at the lottery system that gave ordinary citizens, including women, the opportunity to attend the debates and voice their opinions in an atmosphere that Williams described as simultaneously chaotic and exalting[17].

Williams sought not only to describe the new political institutions in revolutionary France but also to create an interest for her readers in the success of the new government that held promise for the enlightened transformation of all of Europe. In her *Sketches* she dedicated several letters to the Neapolitan revolt, including the following comment: «The authenticated history of Jacobin terror in France, or of royal terror at Naples has traits enough to darken the deepest tragedy, and make us shudder at the reflection of their real existence» (Williams 1801:123). So, Williams presented the Neapolitan revolution as "the consciousness of its feebleness in the struggle[18].

In conclusion, after having briefly presented the well-known Neapolitan revolution, this paper has focused on the revaluation of testimonies by characters who lived through the revolution and perceived it in several ways, from the voices of famous intellectuals to the supporting actors, without neglecting women. Pimentel, Sanfelice and many others devoted their lives to the revolution and the dream of freedom. These women faced distinct challenges as political actors and used strategies to justify their own involvement that connected liberty with passion and privileged feminine sensibility in the revolutionary period. Throughout history they have been often presented as lost and lonely women, but they were virtuous, noble women who were expiated cruelly, becoming protagonists for a day. Helena Maria Williams was not an eyewitness of the Neapolitan revolution, but because of her experience of the French revolution, the Naples revolt influenced and impressed her so much that she dedicated several letters to this tragic Neapolitan period.

Finally, from the chronicles of the Neapolitan revolution and counter revolution, we can once again confirm that the 'populace' played a key

---

17 All the nations of Europe following the liberal system which France has adopted, the little crooked policy of the present times shall give place to the reign of reason, virtue, and science. Ibid., 82.
18 As this treaty on the part of the Court of Naples had been dictated by the consciousness of its feebleness in the struggle, instead of being the result of pacific sentiment, or moderation, it was not difficult to foresee, that its duration would be of short continuance. Ibid., 125.

role in all the accounts. The indifference of the populace was the cause of a feeble revolution and the consequent failure of the republican period. The lack of education of the populace was another decisive aspect – they did not understand the republican message. With the entry of the Sanfedist troops and the return of the Bourbon monarchy, the chronicles do not portray a passive or feeble populace but one that is simply non-reactive, almost deaf to the republican messages and ready to restore the monarchy with cruelty and violence.

Bibliographical references

Abulafia, D., *The great Sea: A Human History of the Mediterranean* (London: Penguin, 2015).
Ajello, R., *Il problema della riforma giudiziaria e legislativa nel Regno di Napoli durante la prima metà del secolo* (Napoli: XVIII Ed.,1961).
Battaglini, M., *La rivoluzione giacobina del 1799 a Napoli* (Messina-Firenze: D'Anna, 1973).
— *Napoli tra monarchia e repubblica. Note e postille* (Roma: Amal, 1996).
— *Eleonora Fonseca Pimentel: il fascino di una donna impegnata fra letteratura e rivoluzione* (Parma: G. Procaccini, 1997).
Benigno, F., *Specchio della Rivoluzione: conflitto e identità politica nell'Europa moderna* (Roma: Donzelli, 1999).
Buttafuoco, A. R., *Eleonora Fonseca Pimentel: una donna nella rivoluzione*, Donna e Ricerca Storica, Roma, 3 (1977).
Casaburi, M., *Fabrizio Ruffo: l'uomo, il cardinale, il condottiero, il politico* (Roma: Rubettino, 2003).
Chiosi, E., 'Il regno dal 1734 al 1799', in G. Galasso and R. Romeo (eds), *Storia del Mezzogiorno* (Napoli: Edizione Del Sole, 1986), Vol. IV, Book 2, pp. 373-467.
Cohen, D., *La nature du peuple. Les formes de l'imaginaire social (XVIIIe-XXIe siècles)* (Paris: Champ Vallon, 2010).
Croce, B., *Studi storici sulla rivoluzione napoletana del 1799* (Roma Trani: E. Loescher tip. V. Vecchi, 1897).
— 'Nel furore della reazione. Dalle memorie di un milite della Guardia civica della Repubblica napoletana', *Archivio storico per le Province Napoletane*, XXIV (1899).
— *La rivoluzione napoletana del 1799. Biografie, racconti, ricerche*, 3th edn (Bari: Laterza, 1912).

— *Luisa Sanfelice e la congiura dei Baccher: racconto storico* (Bari: Laterza, 1942).

— *Prefazione a La riconquista del Regno di Napoli nel 1799; lettere del cardinale Ruffo, del re, della regina e del ministro Acton* (Bari: Laterza, 1943).

— *La rivoluzione napoletana del 1799* (Napoli: Bibliopolis, 1998).

Cuoco, V., *Saggio Storico sulla Rivoluzione di Napoli*, ed. by A. De Francesco (Manduria-Roma- Bari: Lacaita,1998).

D'Ayala, M., *La nobiltà napoletana nel 1799. Vite dei magnifici Cittadini Caracciolo- Carafa-Colonna-Genzano- Doria-Pignatelli- Riario- Serra* (Napoli: Grande Stabilimento Tipografico, 1873).

Davis, J. A., 'Rivolte popolari e controrivoluzione nel Mezzogiorno continentale' in *Le insorgenze popolari nell'Italia rivoluzionaria e napoleonica*, ed. by A. M. Rao, *Studi Storici*, 39 (2) (1998).

De Nicola, C., *Diario napoletano (dicembre 1798- dicembre 1800)*, ed. by P. Ricci (Milano: Giordano,1963).

De Francesco, A., *Rivoluzione e costituzioni; saggi sul democratismo politico nell'Italia napoleonica, 1796-1821* (Napoli: Edizioni scientifiche italiane, 1996).

— *Vincenzo Cuoco. Una vita politica* (Roma-Bari: Laterza, 1997).

De Lorenzo, G., *Nel furore della reazione del 1799*, Intr. and note by B. Croce, ed. by C. Cassani (Napoli: Colonnese, 1998).

— *Memorie*, ed. by P. Russo (Napoli: Vivarium,1999).

De Lorenzo, R., 'La tradizione del 1799 nel Risorgimento italiano', in *Archivio di Stato di Napoli, La Repubblica Napoletana del Novantanove. Memoria e mito* (Napoli: Macchiaroli Editore,1999), pp. 91-110.

De Martino, A. S., *Antico regime e rivoluzione nel Regno di Napoli. Crisi e trasformazione degli ordinamenti giuridici* (Napoli: Jovene, 1971).

D'Episcopo, F., *Eleonora De Fonseca Pimentel tra mito e storia* (Napoli: Edizioni Scientifiche Italiane, 2008).

Di Battista, F., *Il Mezzogiorno alla fine del '700* (Roma Bari: Laterza, 1992).

Di Franco, S., 'Le rivolte del Regno di Napoli 1647-48 nei manoscritti napoletani', *Archivio Storico per le Province Napoletane*, CXXV (2007) 1-132.

Di Franco, S., and A. Musi (eds), *Mondo antico in rivolta (Napoli 1647- 1648)* (Roma-Bari: Lacaita, 2006).

Fonseca Pimentel, E. *Il Monitore Napoletano del 1799. Articoli politici seguiti da scritti vari della stessa autrice*, ed. by B. Croce (Bari: Laterza, 1943).
— *From the Arcadia to the Revolution. The Neapolitan Monitor and other Writings*, ed. and trans. by V. R. Jones (Toronto: Iter Press, 2019).
Forgione, M., *Eleonora Pimentel Fonseca. La straordinaria avventura politica e umana di una protagonista della Repubblica napoletana del 1799* (Roma: Newton Compton, 1999).
Gabrielli, P. (ed.), *Vivere da protagoniste: donne tra politica, cultura e controllo sociale* (Roma: Carocci, 2001).
Hugon, A., *Naples insurgée 1647-1648: De l'évènement a la memoire* (Rennes: Presses Universitaire de Rennes, 2011).
Kennedy, D., *Helen Maria Williams and the Age of Revolution* (Lewisburg: Bucknell UP, 2002).
Lomonaco, F., *Rapporto al cittadino Carnot, preceduto dalla traduzione dei «Droits et devoirs du citoyen» di Gabriel Bonnot Mably*, ed. by A. De Francesco (Roma-Bari: Lacaita, 1999).
Orefice, A., *Eleonora Pimentel Fonseca. Eroina della repubblica napoletana del 1799* (Roma: Salerno Editore, 2019).
Macciocchi, M. A., *Cara Eleonora. Passione e morte della Fonseca-Pimentel nella Rivoluzione napoletana* (Milano: Rizzoli, 1993).
Malanima, P., *La fine del primato. Crisi e riconversione nell'Italia del Seicento* (Milano: Mondadori, 1998).
Mafrici, M., *Il Mezzogiorno settecentesco attraverso i catasti onciari*, 2 vols (Napoli: Edizioni Scientifiche Italiane,1986).
Massafra, A., 'Fisco e baroni nel Regno di Napoli alla fine del secolo XVIII', in *Studi storici in onore di Gabriele Pepe* (Bari: Dedalo,1969).
Matarazzo, P. (ed.), *Catechismi repubblicani (Napoli, 1799)*, intr. by E. Chiosi (Napoli: Vivarium, 1999).
Orefice, A., *La penna e la spada. Particolari inediti su Eleonora de Fonseca Pimentel ed Ettore Carafa conte di Ruvo* (Napoli, Istituto Italiano per gli Studi Filosofici, 2009).
Parker, G., *Global crisis: War, Climate Change and Catastrophe in the Seventeenth Century* (New Haven: Yale University Press, 2013).
Petromasi, D., *Storia della spedizione dell'Eminentissimo Cardinale D. Fabrizio Ruffo allora vicario per S.M. nel Regno di Napoli e degli avvenimenti fatti d'armi accaduti nel riacquisto del medesimo compilata da D. Domenico Petromasi* (Napoli: Vincenzo Manfredi, 1801).

— *Alla riconquista del Regno di Napoli. La marcia del cardinale Ruffo dalle Calabrie a Napoli*, intr. by S. Vitale (Napoli: Il giglio, 1994).
Pellizzari, M. R., *Eleonora de Fonseca Pimentel: morire per la rivoluzione*, Storia delle Donne, 2008.
Pietrabondio, D., *Anarchia popolare a Napoli*, ed. by M. Arcella (Napoli: De Angelis, 1884).
Rao, A. M., *Il regno di Napoli nel settecento* (Napoli: Guida, 1983).
— 'Napoli e la rivoluzione (1789-1794)', in *"Prospettive settanta"*, 7 (1985), pp. 403-476.
— 'La Repubblica Napoletana del 1799', in *Napoli e la Repubblica del '99. Immagini della rivoluzione, catalogo della mostra, Napoli, Castel Sant'Elmo, 13 dicembre 1989-28 gennaio 1990* (Napoli: Elio De Rosa, 1989).
— *Esuli. L'immigrazione politica italiana in Francia (1799-1801)* (Napoli: Guida, 1992).
— 'Tra riforma e rivoluzione: Michele Torcia (1736-1807)', in P. Macry and A. Massafra, *Fra storia e storiografia. Scritti in onore di Pasquale Villani* (eds) (Bologna: Il Mulino, 1994).
— *Napoli 1799 fra storia e storiografia* (Napoli: Viarum, 2002).
— 'Eleonora De Fonseca Pimentel, le monitore napoletano et le problème de la partecipation politique', *Annales historiques de la Rèvolution Française*, 344 (2006).
Romano, R. *Opposte congiunture. La crisi del Seicento in America e in Europa* (Venezia: Marsilio, 1992).
Ruffo, G. and D. De Maio, *Il cardinale Fabrizio Ruffo tra psicologia e storia: l'uomo, il politico, il sanfedista* (Roma: Rubettino, 1999).
Sacchinelli, D., *Memorie storiche sulla vita del Cardinale Ruffo* (Napoli: Cataneo, 1836).
Santos, M. T., 'Leonor da Fonseca Pimentel. A Portuguesa de Nápoles (1752-1799)' in *Actas do colóquio realizado no bicentenário da morte de Leonor da Fonseca Pimentel* (Lisbona: Sara Marques Pereira, 2001).
Schiattarella, F., *Marchesa giacobina: Eleonora Fonseca Pimentel* (Napoli: Schettini, 1973)
Stone, B., *Rethinking revolutionary change in Europe: A Neostructuralist Approach* (Lanham- Boulder-NewYork London: Rowman & Lietfield, 2020).
Urgnani, E., *La vicenda letteraria e politica di Eleonora De Fonseca Pimentel* (Reggio Calabria: La Città del Sole, 1998).

Villari, R., *Un sogno di libertà. Napoli nel declino di un impero. 1585-1648* (Milano: Mondadori, 2012).

Vovelle, M. and J. Davis, 'The Neapolitan Revolution of 1799. Popular societies in the Neapolitan republic of 1799', *Journal of Modern Italian Studies*, 4 (1999).

Williams, H. M., *Letters Written in France, in the Summer 1790, to a Friend in England; Containing, Various Anecdotes Relative to the French Revolution; and Memoirs of Mons. and Madame du F* (London: Thomas Cadell, 1790).

— *Letters From France; Containing a Great Variety of Original Information Concerning the Most Important Events that Have Occurred in that Country in the Years 1790, 1791, 1792, and 1793* (Dublin: J. Chambers ,1794).

— *Letters containing a sketch of the politics of France from the thirty-first of May 1793 till the twenty eight of July 1794 and of the scenes which have passed in the prisons of Paris, vol. IV* (Dublin: J. Chambers, 1796).

— *Sketches of the State Manners and Opinions in the French Republic towards the close of eighteenth century in a series of Letters* (London: G. G. and J. Robinson, 1801).

SILVANA RAFFAELE

11.
A WAR BASED ON SLANDER, TRICKS AND QUIBBLES. THE CONFLICT BETWEEN HOMEOPATHY AND ALLOPATHY IN THE BOURBON SOUTH

The world of health and medicine, with the myriad problems that have characterized it over time, includes many aspects of a civilized way of life, which can be traced back to the connections in the mosaic of the social sciences. This allows us to open the subject to wider re-interpretations – in socio-anthropological, economic and political terms – regarding a community in a precise historical moment.

The analysis of the history of medicine is closely connected to human suffering, but also to the power of those who exercise control over health, the places assigned to it, and the institutions, which between the 18th and 19th centuries – the time frame examined here – also played an important role in the field of medicine.

The age of Enlightenment triggered important changes, and, in the wake of Cabanis, was also to influence ideas concerning medicine, consolidating, as in politics, the opposition to the dogmatism of the ancient regime and the openness to new demands.[1]

The old dispute between science, alchemy, abstractionism and dogmatism had been resolved, and this mixed bag of problems was to pave the way for new polemics and conflicts. The new doctors and the surgeons – who had earlier been compared to barbers – eclipsed the old doctors/philosophers, even usurping their places of power in urban societies, along with those of old classes of noblemen and dignitaries. Thus, ancient alliances were disrupted at the same time as the new political-institutional and socio-cultural changes took place that characterized the formation of the modern State. Scientists and intellectuals, advocating new systems, were fiercely opposed by the old generations, but welcomed with enthusiasm by young people who were also politically engaged.

In the medical field in particular, a renewed holistic vision was taken up that focused not so much on getting rid of the disease, but on treating the patient as a whole, in spirit and body (Rossi 2012: 32). In past times,

---

1   cf. Raffaele 2008; Moravia 1974; Cosmacini 1988; Frasca 2008.

the Pythagorean school had considered the human body as a harmonious microcosm that responds to disequilibrium with the symptom "disease". Even Hippocrates, the first to express the famous saying *similia similibus curantur*, and Galeno after him, saw disease as an imbalance to be treated with the rebalancing of moods. The alchemist doctor Paracelsus (Vecchio et al. 2011: 153-161) later experimented the Hippocratic assumption in the field. Even Kepler and Newton succumbed to the charm of alchemy and astrology while studying chemistry, physics, anatomy and physiology, and encouraging experimentation (Frasca 2014: 15).

Homeopathy[2] one of these new theories, can be considered as a paradigmatic element of the topic in question; it was included in a web of disputes that brought into play several conflicting factors: old and new doctors, aromatics, institutions and politicians, innovators and traditionalists.

Christian Friedrich Samuel Hahnemann (Bradford 1993), challenged by the strict traditionalists, but followed – along with Brown – by new generations hoping for epochal changes, was one of those who at that time made theoretical proposals concerning the idea of the homeopathic method, in harmony with vitalism, as opposed to the ingrained allopathy. With the new term recalling Hippocrates, Hahnemann wanted to highlight the philosophical difference at the base of the two approaches: a mechanistic one that focuses attention on the "sick organ in the case of allopathy, and the dynamic approach of homeopathy that takes the whole man into consideration. Homeopathy tends to focus on "vital energy", considering all the physical and psychological aspects of the individual within a holistic approach: the symptom thus becomes a signal to be interpreted, starting from the dialogue between doctor and patient, to identify the reactions of each individual patient.

While translating the *Materia Medica* by William Cullen (Frasca 2014: 13), Hahnemann took note of the efficacy of quinine. He refined the system and decided to experiment the medicine on himself. After taking it for several days, he felt he had developed the signs of malaria; he hypothesized that a series of symptoms could be treated with that substance, in a strongly diluted form, that would produce the same effects in a healthy person. In 1810 he published his experiments in the first edition of his Organon (Hahneman 2003).

---

2   cf. Brambilla 1984: 4-15; Lodispoto 1987; Negro 1996; Charette 1998; Armocida and Zanobio 2002.

In 1812 in Lipsia, where he had moved from his native Meissen after various events, and where he was affiliated with Freemasonry (Francovich 1974), he founded a school of homeopathy, which met the strong opposition of doctors and pharmacists. The story of Prince Schwarzenberg, who was entrusted to Hahnemann's care and later died, heightened the conflict between proponents and opposers of the system. However, it did not prevent the spread of his theory (Gazzetta di Firenze 1820; Giarrizzo 1994; Landolina 2006).

Homeopathy arrived in Italy with the Austrian troops who were called in after the revolts of 1821 (Romeo 1950). Many of the army's military doctors, such as Merenzeller, the chief physician, already officially practiced this system. General Köller had even donated Hahnemann's *Organon* and *Materia* to the Royal Academy of Sciences. In response, the Academy sent the Austrian physician Schoenberg to practice with Hahnemann; Schoenberg later published *Il sistema medico del Dottor Samuele Hahnemann esposto alla Reale Accademia delle Scienze di Napoli* (von Schoenberg 1822; Panvini 1824; Quaranta 1824) ("The medical system of Dr. Samuele Hahnemann explained at the Royal Academy of Sciences in Naples"). The homeopathic theory was easily accepted in Naples, where the widespread diffusion of masonry had found fertile ground due to military lodges in the Austrian army stationed in the Kingdom.

The protection of Maria Carolina and Acton (Recca 2014) gave rise to a new flowering of the phenomenon which, however, ground to a sudden halt with the French revolution and the beheading of Marie Antoinette. This provoked a strongly anti-masononic sentiment in the court with the consequent transfer of the surviving "brothers" to the Neapolitan patriotic lodge, which was to favor the subtle and dynamic masonic work by Carlo Laubergh (Croce 1989: 363-437). The Bourbon Ferdinand I can be placed in this framework in some way; he was a sympathizer of the Saint-Simonianism movement (as can be seen by the experiment of S. Leucio), and then of the Napoleonic model – also adopted by his restored French cousins – of which he would leave strong traces in his five-year reforms (Raffaele 2005). Finally, the Austrian troops were called back due to the upheavals of the Italian Unification, demonstrating the link between the two revolutions – medical and political – and facilitating a sort of triumph of homeopathy amid political and ideological conflicts, which reached a culmination when Francesco I, cured of a severe angina pectoris by this method, took the initiative to promote the spread of the new theory. There were two well-known successful doctors at the Bourbon court in Naples: Cosmo De Horatiis (De Horatiis 1845) – director of the university surgical clinic and a student of Giorgio Necker di Melnik (Lodispoto 1987: 73) – who opened a dispensary of homeopathic medicine in the capital of Naples and directed the first specialist journal in 1829 *Effemeridi di medicina omiopatica* ("The Ephemeris of Homeopathic Medicine"), and Francesco Romani.[3]

The extraordinary event of the healing of an eye tumor suffered by Field Marshal Joseph Franz Karl Radetzky von Radetz, which even the famous Austrian court ophthalmologist Friedrich Jäger (Negro 2005: 36-38), expressly sent by the emperor, had given up treating, definitely marked the success of homeopathy in Italy (Piterà 1998: 25-31). Jan Christoph Härtung (Lodispoto 1987: 121), a Saxon homeopathic doctor from Salzburg hospital, radically cured the general in just six weeks. When a member of the Royal Academy of Medicine in Brussels asked for confirmation, Radetzky (Granier 1855: 91) replied with this letter:

> Sir, it is with pleasure and gratitude that I declare that it is to Hartung, a homeopathic doctor, that I owe the healing of a very serious disease of the eye. Finding myself already abandoned by the other doctors, it is thanks to Homeopathy that I have sight, as well as life. Signed: Radetzky.

The event was "forgotten" in Italy, for reasons we can define as patriotic, because it was linked to the hated Austria. However, in Germany and throughout Europe the Radetzky case was widely reported. Homeopathy spread in Sicily through doctors, one of the first being Giuseppe Tranchina (Lodispoto 1987: 61-62), who spread it in Sicily in 1829. According to Paolo Morelli (Morello 1838; Id. 1847; Sampolo 1873). Tranchina had practiced it after learning it in Naples in 1821 from Barthlè, a military doctor of the Austrian army.

So it was that in 1829 the *Effemeridi di Medicina Omiopatica* ("The Effemeris of Homeopathic Medicine") was founded, directed by Cosmo De Horatiis. In 1831 *L'Hannemanni, Giornale Omeopatico di Sicilia* ("The Hahnemann Journal of Homeopathic Medicine in Sicily") was published, edited by Cavallaro, and the *Annali di Medicina Omiopatica* ("Annals of Homeopathic Medicine"). In 1855, the first issue of the magazine *Giornale Scientifico dell'Accademia Omiopatica di Palermo* (Negro and Negro, 2007: 99) ("The Scientific Journal of Homeopathic Medicine") came out.

In 1862, Morelli, assisted by Cataldo Cavallaro (Cavallaro 1845; Id. 1847; Id. 1871), wrote the following in the first issue of the *Annali di*

---

[3] cf. Romani, 1816; Id. 1837; Id. 1845; Pietrocola 1997; Hahnemann 2003.

medicina omiopatica di Palermo, recalling his ideological and professional transition at the age of 36 in 1831, from allopathy to homeopathy:

> When Homeopathy was introduced in Sicily, mainly thanks to Dr. De Blasij Antonino, Bartoli Andrea and Mure Benedetto ... they began to build it on two cornerstones: the first one was the free dispensary; the second was the *Annali di Omeopatia per la Sicilia* (Annals of homeopathy for Sicily) ... the third stone to be placed was the *Accademia Omeopatica Sicula* (Sicilian Academy of Homeopathy). We trust that Bartoli, whose respectable and vigorous old age is evidence that he was always a master in the study of our medicine, will be generous in communicating the fruits.

Morello had also been indoctrinated by another significant figure: Michele Foderà (Grillo 2011). He had trained in Palermo, Paris, and once again in Palermo, in democratic and masonic circles, and is likely to have approached Utopianism and, in 1840, socialism. Therefore, having converted to homeopathy, Morello collaborated with the *Annali di Medicina omeopatica*, ("Annals of Homeopathic Medicine"), together with Antonio De Blasi (De Blasi n.d.) – who was also close, along with Giuseppe Bandiera (Bandiera 1842; Id. 1845), to the Fourierist circles in Palermo – and created in the Sicilian capital, the *Accademia di Medicina omeopatica* (*Statuti* 1852) (the Academy of homeopathic Medicine), connected to the Fourierist circle. In this direction, as Comte and Cabanis had suggested, the purpose of studying biology widened from the understanding of man to the system of society, underlining the link between nature and the social order. This group was headed by Benoît Jules Mure, who in Palermo would promote homeopathy, considering it in tune with the Fourier thought which he had learned about through his friend – and the successor of Fourier himself – Victor Considérant (Chironna 2016).

The young Mure, born in Lyon to Italian parents, had long been suffering from tuberculosis, and had come to Palermo in 1833 in search of a more favorable climate. Already on his arrival in the island, in Messina, he suffered a severe episode of hemoptysis that was homeopathically halted by Marshal Luigi Carafa of the Princes of Noia (Carafa di Noja 1838), who advised him to return home, where he was treated by Sebastiano De' Guidi (De Guidi 1836), pupil of Romani; after a few months he was able to return to Sicily. When he arrived in Palermo, he became the main instigator of the diffusion of the new homeopathic theory, with the help of Giuseppe Tranchina, Antonio De Blasi, Samuele Calandra (Chironna 2016: 32), and the abbots Biagio Tripi and Antonio Bandiera. The latter had already started publishing the *Annali di medicina omeopatica* ("Annals of homeopathic medicine"), a periodical in the 1830s that would carry on until 1845 (*Annali* 1837).

In 1836, De Blasi had published the *Avviso al popolo sul trattamento del Cholera Morbus* (De Blasi 1836), ("Notice to the people on the treatment of Cholera Morbus"), arousing the hostility of the medical authorities. The bitter conflict that ensued caused his expulsion from the Palermo Academy of Sciences of which he was secretary.

In 1837, cholera (Tripi 1837; Del Panta 1980; Tognotti 2000) spread like wildfire amid controversy, anti-bourgeois accusations and open rebellion: a large percentage of the cases treated by De Blasi were cured, and between the end of 1837 and 1838 Mure opened the first *Dispensario omeopatico* ("Homeopathic Dispensary") in Palermo.

From 1838 the hospital *Fratelli di San Giovanni di Dio* also adopted homeopathic treatments under the direction of Antonio Bandiera, who was close to Mure. Given the religious nature of the institution, no information was given as to what was happening in the wards; as a result, no controversy intruded on the silent life of the hospital and its doctors.

Meanwhile, Mure was preparing the equipment to simplify the preparation technique of homeopathic medicines so that he would not need to import them. He also undertook an advertising campaign by bombarding the city with placards to publicize the opening of the dispensary where Biagio Tripi, Antonio De Blasi and Samuele Calandra treated with great self-sacrifice the patients who crowded the sanatorium.

In May 1839 Mure left Sicily.[4] His departure and, later, that of Calandra, had a drastic effect on the dispensary. With Mure's departure from Sicily, due to one conflict after another, the dispensary he had created was finally closed and the homeopathic center destroyed: small rivalries, personal or school ambitions had prevailed. After the establishment of a new homeopathic dispensary in the same year (financed by a benefactor), Antonio De Blasi and Paolo Morello formulated a statute in 1840, asking the government to establish the *Accademia omeopatica* ("homeopathic Academy"). The resentment of the conservative doctors of the Royal Academy of Sciences succeeded at first in blocking the initiative. However, following an aggressive press campaign, Paolo Morello wrote: *Considerazioni sul rapporto del collegio Medico contro la domanda degli omeopatici in Sicilia* ("Considerations on the report of the Medical College against the demand of homeopaths in Sicily"), and the *Commissione della*

---

[4] Afterwards, Mure spread the theory and practice of homeopathy in France, Brazil, Egypt and the Sudan.

Pubblica Istruzione (the Commission of Public Instruction) was finally forced to give a favorable opinion (Morello 1862; Id. 1865).

Moreover, Ferdinand II unexpectedly granted the founding of a Dispensary and an Academy of Homeopathic Medicine as recognition of the work carried out by doctors during the cholera epidemic of 1837 in Sicily, the statute of which, however, was to be supervised by members of the *Accademia allopatica* (Allopathic Academy) with right of veto. Numerous quibbles and various hitches made the process long and difficult, but on 23 June 1844 Andrea Bartoli, as the first president of the Academy, could finally inaugurate its opening on 23 June 1844. Three years later the first diplomas of Doctor in Homeopathic Medicine were awarded (Lodispoto 1987: 68-69).

Meanwhile, homeopathy spread to the rest of the island; it arrived in Catania in 1847 with Sebastiano Cappellani, where the doctor from Palazzolo asked, in vain, for permission to hold a university course of homeopathy in the University. On April 13, 1850 the Medical College communicated its refusal. The government, with a ministerial note dated October 19, 1850, solicited the opinion of the Palermo Homeopathic Academy which, obviously, granted their approval the following January.

At the same time, in Palermo, Tripi.[5] published a *Confutazione al rapporto del Collegio Medico di Catani* ("Refutation to the report of the Medical Board of Catania") that received no official answer. The fierce controversy was also reported in the scientific journals of the island. Ignazio Ficuccia, a young student from Catania University, published a particularly significant article in the *Giornale del Gabinetto Letterario della Accademia Gioenia* ("Journal of the Literary Cabinet of the Gioenia Academy") titled *Poche osservazioni su un opuscolo stampato a Palermo* ("A few observations on a pamphlet printed in Palermo"), referring to what Tripi had written in the *Hahnemann* of 31 March 1847 in an unsigned article. We thus became aware of new details regarding the beginnings of homeopathy in Catania (Lodispoto 1987: 74):

> [...] in 1840 the Royal Receiver of the Customs of Catania, having returned to Palermo, brought from that capital a box of homeopathic liquids that Dr Mure had given him, plus the starchy sugar confections and the Annuals of Dr. De Blasi, which were read, liked and were also purchased by Mr. D. Salvatore Mancino, by the mathematical professor Mr. D. Gaspare Gambino, and others; and they all provided themselves with medicines, and although they

---

5 cf. Tripi 1838; Id. 1842; Id. 1844a, Id. 1844b; Id. 1847; Id. 1852; Id. 1855; Tebaldi 1847.

were laymen, managed to produce wonderful treatments followed by sound healing. In 1845, the surgeon Dr. Giuseppe Politini from Palagonia came to Catania where he performed amazing healings, but he left before 1847 and homeopathy returned to the hands of the laity, among whom the aforementioned mathematician Gambino of very high birth distinguished himself, and he continued in his philanthropic homeopathic exercise until 1849, and helped people if they were sick. In March 1847, Dr. Cappellani came for several days for professional matters and was visited morning and evening by the famous professor of philosophical chemistry Sir Carmine Maravigna, who provided at that time the library of this Royal University of the Organon of Hahnemann ...; and he, Dr. Maravigna, visited the sick, who he treated, already of course, homeopathically.

Despite the established knowledge of some fixed points, the spread of Hahnemann's theory sparked parochial squabbles for the attribution of the primacy of the introduction of homeopathy on the island. Trapani, for example, deliberately ignoring the events in Palermo, proclaimed that the method had been introduced in 1845 by a bookseller, a certain Pietro Colajanni, who had been seriously ill and, after trying every allopathic cure possible, had gone, coincidentally, to his native Palermo, to consult the masters of homeopathy, Bartoli and De Blasi, who had cured him. Then he began to buy books and medicines, while encouraging the diffusion of the new method (Pompili 1864: 182; Lodispoto 1987: 88).

Essays on homeopathy appeared on the shelves of the bookshops of the island, while fierce arguments broke out between allopathic and homeopathic pharmacists.

A new wave of cholera in 1854 brought the conflict between old and new medical systems back into the limelight. When the epidemic began, the Sicilian lieutenant, the prince of Satriano, encouraged by the Neapolitan successes of Biagio Tripi, entrusted him with the health of the troops and the military hospital of S. Cita (Tripi 1856) in Palermo. In 1855, also in Palermo, a periodical was started titled *Giornale Scientifico dell'Accademia Omiopatica* ("Scientific Journal of the Homeopathic Academy"), edited by Tripi himself.

A war of shrewdness, spite, slander, obstacles and impediments was rekindled that led to the closure of the clinic that he directed, and soon afterwards also the journal.

Gioacchino Pompili wrote (Pompili 1864):

the results were very happy, the rage of the allopaths immense, and the war fierce. The slanders were endless. In view of this only, after six months of successful experiments, Dr. Tripi did not want to continue for the sake of peace and quiet for the clinic, which was closed with great honor on the part of homeopathy and with new evidence of the bad faith of allopaths.

However, the homeopathy phenomenon spread out radially, also investing unusual fields.

The story of the physician Giuseppe Migneco and the poet Mariannina Coffa seems emblematic of the intertwining of homeopathy, masonry, magic and even poetry.

At the end of the 1850s, Migneco – who had graduated in Catania in 1842 and perfected his studies in Naples – was already an important figure in the field of homeopathy, having been very successful during the epidemic of Asian cholera when faced with the objective uselessness of the remedies of allopathic medicine.

He had also overcome the grave injustice made against him by the authorities, who on August 10 1858 burned his monumental work of seven books on hygienic and pathological physiology, following the denunciation of the parish priest of Vizzini that he "exercised diabolic art" (Fiume 2013: 63).

The fateful meeting with Coffa, «poetess of the fatherland and of pain» (Leanti 1923), who had long been afflicted by an undiagnosed organic disease that caused her considerable bleeding, marked his success.

Coffa, who had moved to Noto, blindly relied on a mixed therapy of homeopathy, Mesmerism (Thuillier 1996; Traetta 2007), hypnosis and magnetism, with shades of magic and spiritualism, practiced by Lucio Bonfanti under the direction of Migneco himself. The poetess diligently followed the prescriptions of these doctors and cited them with expressions of great esteem in the poems *Psiche* (*Psyche*), and *Luce e Tenebre* (*Light and Shadow*).[6] All this prompted her husband Morana Frasca to takeswift, serious legal action.

The Coffa case demonstrates in an exemplary manner the presence of a composite environment in Noto where there were different currents of thought: some scientific, some spiritualistic, some others holistic, which aimed at looking at a person as a whole.

The story of Giuseppe Migneco and Mariannina Coffa introduces us to the convoluted paths of sleepwalking, the arcane world of animal magnetism (Amadou 1971), and Mesmerism – systems that were anathema to the Pope but were cultivated within democratic masonic élites. Coffa's

---

6   cf. Coffa Caruso 1876; Id. 1955; Piterà 1878; Bonfanti 1879; Coffa 1879; Sbano 1879; Di Stefano 1996; Fiume 2000.

poem dedicated to Giulia Caico, who she had met through their mutual friend Migneco and with whom she had an epistolary relationship, was certainly inspired by masonic and magnetic-spiritistic concepts. Giulia was the sister of Cesare and Federico, the Princes Caico of Montedoro, well known, though opposed, local masons. Migneco was especially friends with Cesare and he was a long-time guest at the Caico home (Gernia 1872; Migneco 1855). Giulia Caico was also treated by Mignecoat his home in Catania.

The prince also had several of the many works of his physician friend printed at his own expense, and he supported him in various court cases against his enemies.

In 1865 cholera broke out in Montedoro. And while the controversy regarding homeopathic treatments intensified, once again Cesare Caico invited Giuseppe Migneco to his home; he also published *Appello ai cento comuni e popoli d'Italia, per la nuova alleanza di salute*, (*Appeal to the hundred municipalities and peoples of Italy, for the new health alliance*) and made a request for contributions for the preparation of medicines.

On March 16, 1867, Migneco received a certificate of public esteem and gratitude from the City Council of Montedoro (the mayor was Onofrio Caico) for having protected the citizens from the cholera epidemic.

Finally, on June 9 the city council awarded him a diploma and a medal, with the endorsement, from Florence, of the Minister Secretary of State for Internal Affairs.

Another interesting case has the elements of a thriller. Michele Pappalardo (Messana 2007; Fiume 2000: 33-47, 129), a homeopathic doctor-pharmacist, mysteriously disappeared from the train to Rome. His family members said that he had with him a small bottle of Migneco's famous remedy against cholera, to be deposited in the appropriate offices to obtain a possible patent. Pappalardo had also strenuously defended his "master" Migneco in a booklet of a few pages, making himself unpopular in the eyes of the many doctors who saw him as a dangerous rival (Pappalardo 1868). He was taken in a carriage from Montedoro to Serradifalco station and was never seen again.

In April 1868, in Palagonia, Giuseppe Politini Trayna, scientist and doctor of well-deserved fame, who had introduced homeopathy in Catania, was suffering from a serious illness and was being treated by his nephew and aspiring heir, Dr. Giuseppe Politini Vecchio (Politini Vecchio 1885).

On April 17, Migneco received some news that prompted him to visit his friend. He predicted a fatal prognosis if he did not intervene immediately with the appropriate homeopathic treatment. The nephew bitterly refuted the diagnosis; indignant and disappointed, he abandoned the sick man and published a pamphlet against Migneco, calling him *Piccolo Cagliostro* (Politini Vecchio 1868), The patient registered a marked improvement with the suggested homeopathic treatments, but the nephew pressed him to abandon them. Giuseppe Politini Trayna died, mourned by Migneco, who rushed to his bedside to say goodbye.

The story rekindled the conflict between old and new systems. It was the year 1870, and the controversy was fresh. The derogatory text against Migneco signed by Politini Vecchio testifies to the atmosphere: even an 'alleged' healing by homeopathic treatment, considered to be demonic magic, was denied in order to demonstrate the supremacy of allopathic medicine.

Although homeopathy had won the hearts of the ruling class, an act was required to sanction and affirm its importance: admission to the university. As we have seen, the Academy had obtained the right to teach it and was able to issue a doctor's diploma in homeopathy. However, this could not be compared to the special aura of the dignity and importance of a university chair. On February 12, 1862, Dr. Paolo Morello, on behalf of the Homoeopathic Academy of which he was president, addressed a petition to the Italian Parliament for the theoretical and practical teaching of homeopathy.

His request was ignored; times had changed. 1837 had already marked the definitive detachment between the regime and the island, marking the divorce between Sicilian democracy and the Bourbon dynasty, and in any case the Bourbons were in exile. There was no longer the support that for many years had protected the followers of the theories of Hahnemann that had found their first and most significant following in the Kingdom of the Two Sicilies.

However, the validity of homeopathy had been established: although it was seen with some suspicion, a considerable number of doctors became interested in it in those crucial years. The conflicts, as we have seen, were neither few in number nor small in significance. The problem started from a medical dispute, but in reality, it became, together with the contemporary Brunonian system of medicine, a sort of sounding box of socio-cultural and political-economic matters of far wider scope.

> The exchange of illuminated ministers *al di qua del Faro* (referring to southern Italy as far as the lighthouse of Messina) and the contemporary turn-over of viceroy innovators *al di là del Faro* (beyond the lighthouse of Messina, Sicily) reflect the image of a southern Italy animated by old barons and new

philosophers, between old reluctance and modern expectations" (Frasca 2008) and looking for a more modern but not easy equilibrium.

The homeopathic metod – be it alternative or 'heretical' – which merges philosophy, science and politics, had and still has conflicting successes; it has been considered beneficial and not harmful by its advocates, and useless and dangerous by its opposers.
But that is another story ... or maybe not.

Bibliographical references

Amadou, R. (ed.), *Le magnetisme animal* (Paris: Payot, 1971).
Annali di medicina omiopatica per la Sicilia, 1, Palermo (1837).
Armocida, G., and B. Zanobio, *Storia della medicina* (Milano: Masson, 2002);
Bandiera, G., *Esposizione della omiopatia, sue vicende e suoi progressi, discorso letto nell'Accademia Omiopatica di cui era vice-segretario* (Palermo: s.e., 1845).
— *Istruzione sulla pratica omiopatica di Samuel Hahnemann* (Palermo: s.e., 1842).
Bonfanti, L., *Lettera di Mariannina Coffa a suo fratello Vincenzo* (Noto: Tipografia di F. Zammit, 1879).
Bradford, T., *La nascita dell'omeopatia. Vita e lettere di Samuel Hahnemann* (Milano: Perla edizioni, 1993).
Brambilla, E., 'La medicina del Settecento', in F., Della Peruta (ed.), *Storia d'Italia, Annali 7, Malattia e medicina* (Torino: Einaudi, 1984), pp. 4-15.
Carafa di Noja, L., *Lettera a G.B. Mure*, Annali di Medicina omiopatica per la Sicilia, 3 (Palermo: Tipografia Spampinato, 1838).
Cavallaro, C., *Medicina omeopatica* (Palermo: s.e., 1845a).
— *Trattato omiopatico sulle malattie della pelle* (Palermo: s.e., 1845b).
— *L'omiopatista un viaggio* (Palermo: s.e., 1847).
— *Corso teorico-pratico-alfabetico di medicina omeopatica* (Palermo: Stamperia di Antonino Russitano, 1871).
Charette, G., *Compendio di omeopatia: la materia medica spiegata* (Palermo: IPSA, 1998).
Chironna, M., *Medici o ciarlatani? L'omeopatia nel Regno delle Due Sicilie dal 1822 al 1860* (Milano: FrancoAngeli, 2016).

Coffa Caruso, M., 'Odea Giuseppe Migneco', in *La donna e la famiglia*. *Versi inediti di Marianna Coffa Caruso in Morana da Noto, pubblicati per cura dell'affezionato ammiratore F. Santocanale* (Palermo: Stabilimento Tipografia Lao, 1876).

— *Poesie in differenti metri* (Siracusa: Stamperia Pulejo, 1955); Fernando Piterà, *Sopra un caso d'isterismo acuto con estasi e sognazione spontanea accaduto in persona della insigne poetessa Marianna Coffa Caruso in Morana. Considerazioni medico filosofiche* (Ragusa: Tipografia Piccitto e Antoci, 1878).

Coffa, V., *Lamento dell'anima a mia sorella Marianna Coffa. Versi* (Noto: Tipografia di F. Zammit, 1879).

Cosmacini, G., *Storia della medicina e della sanità in Italia. Dalla peste europea alla guerra mondiale.1348-1918* (Roma-Bari: Laterza, 1988).

Croce, B., 'La vita di un rivoluzionario: Carlo Laubergh', in B. Croce, *Vite di avventura, di fede e di passioni* (Bari: Laterza, 1989), pp. 363-437.

De Blasi, A., *Avviso al popolo, ossia Notizie sul trattamento omiopatico del cholera morbus* (Palermo: Tipografia del Giornale Letterario, 1836).

— *La mia conversione, discorso sull'omiopatia letta all'Accademia di essa* (Palermo: s.e., s.d.).

De Horatiis, C. M., *Istruzione sulla vaccinazione adottata in Napoli* (Bari: Stabilimento tipografico di Tiberio Pansini, 1845).

de' Guidi, S., *Lettres aux medicins francais sur l'homeopathie* (Lyon: Maison de Commerce, 1836).

Del Panta, L., *Le epidemie nella demografia storica italiana (secc. XIV – XX)* (Torino: Loecher, 1980).

Di Stefano, M. (ed.), *Scritti inediti e rari di Mariannina Coffa* (Noto: Arti grafiche San Corrado 1996).

Fiume, M., *Sibilla arcana, Marianna Coffa (1841-78)* (Caltanissetta: Edizioni Lussografia, 2000).

— *Sicilia esoterica: una guida preziosa per un viaggio iniziatico tra le tenebre dell'isola del sole* (Roma: Newton Compton editori, 2013).

Francovich, C., *Storia della massoneria in Italia dalle origini alla Rivoluzione Francese* (Firenze: La Nuova Italia, 1974).

Frasca, E., *Il bisturi e la toga. Università e potere urbano: il ruolo del medico (secoli XVIII-XIX)* (Bonanno: Acireale – Roma, 2008).

— *L'eco di Brown, teorie mediche e prassi politiche (secoli XVIII-XIX)* (Roma: Carocci, 2014).

Gazzetta di Firenze, 116, Florence, Tuesday 26 September 1820.

Gernia, G., *Cenni biografici dell'illustrissimo avv. Cesare Caico: dei principi di Montedoro* (s.l.: Tipografia Gernia, 1872).
Giarrizzo, G., *Massoneria e Illuminismo nell'Europa del Settecento* (Venezia: Marsilio, 1994).
Granier, M., *Conférence sur l'Homoeopathie* (Paris: Ballière, 1855).
Grillo, M., *Michele Foderà, Un profeta filosofo nella Palermo borbonica* (Catania: Edizioni del Prisma, 2011).
Hahnemann, S., *Organon dell'arte del guarire. Traduzione di Giuseppe Riccamboni dalla 6° ed. tedesca* (Napoli: Cemon, 2003).
Landolina, F., *Logge siciliane tra '700 e '800* (Catania: Centro Grafico, 2006).
Leanti, G., *Una poetessa della patria e del dolore: Mariannina Coffa Caruso (1841-1878)* (Noto: Zammit, 1923).
Lodispoto, A., *Storia della Omeopatia in Italia. Storia antica di una terapia moderna* (Roma: Edizioni Mediterranee, 1987).
Messana, F., 'Pappalardo e Migneco', *La Sicilia*, Catania (2007).
Migneco, G., *Patologia pratica ovvero elementi di clinica omeopatica di Giuseppe Migneco* (Roma: Tipografia Tiberina, 1855).
Moravia, S. (ed.), *La certezza della medicina. Cabanis* (Bari: Laterza 1974).
Morello, P., *Esame dell'Organo della medicina di S. Hahnemann* (Palermo: Tipografia del Giornale letterario, 1838).
— 'Preliminare', *Annali di medicina omeopatica*, I, Palermo (1862).
— *Scienza e libertà* (Palermo: Tipografia del Giornale di Sicilia, 1865).
— 'Saggio di conciliazione tra l'allopatia e l'omeopatia', *Annali dell'Accademia omeopatica siciliana*, (1847).
Negro, A., and F. E. Negro, *Bibliografia omeopatica italiana* (Milano: FrancoAngeli, 2007).
Negro, F. E., *L'Omeopatia* (Roma: Newton Compton, 1996).
— *Grandi a piccole dosi. La parentesi omeopatica di vite famose* (Milano: Tipomonza, 2005).
Panvini, P., *Riflessioni critiche sul sistema medico di Hahnemann esposto dal cavaliere dottor Alberto de' Schoenberg* (Napoli: Stamperia dei fratelli Fernandes e compagni, 1824).
Pappalardo, M., *Brevi considerazioni in occasione del 'Piccolo Cagliostro' di Giuseppe Politini Vecchio* (Montedoro: Stamperia Provinciale Commerciale, 1868).

Pietrocola, G., *Francesco Romani:1785-1852: il medico vastese che introdusse l'omeopatia in Italia ed in Inghilterra* (Vasto: Edizioni del Club Amici di Vasto, 1997).
Piterà, F., 'Il Caso Radetzky. Storia di una Guarigione Omeopatica', *Anthropos & Iatria, Rivista Italiana di Studi e Ricerche sulle Medicine Antropologiche*, II, 2-3 (March-June 1998), 25-31.
Politini Vecchio, G., *Cagliostro il piccolo, storia del dott. Migneco poche parole* (Catania: Coco edizioni, 1868).
— *Il congresso medico internazionale e il cholera morbus* (Catania: F. Martinez, 1885).
Pompili, G., *Il Maresciallo Duca di Saldanha e l'Antiomiopatia o difesa espositiva della dottrina di Hahnemann, in risposta al Discorso di anonimo Professore* (Roma: Menicanti, 1864).
Quaranta, B., *L'Organo della medicina del dott. Samuele Hahnemann trasportato per la prima volta dalla tedesca all'italiana favella da Bernardo Quaranta* (Napoli: dalla tipografia dell'Osservatorio Medico, 1824).
Raffaele, S., *La bottega dei saperi. Politica scolastica, percorsi formativi, dinamiche sociali nel Meridione borbonico* (Acireale – Roma: Bonanno, 2005).
— *Medici e Ateneo: l'onda lunga del potere* (Catania: Cuecm, 2008).
Recca, C., *Sentimenti e politica. Il diario inedito della regina Maria Carolina di Napoli (1781–1785)* (Milano: FrancoAngeli, 2014).
Romani F., *Ricordi sulla peste redatti in un sistema teorico pratico da F. Romani* (Napoli: da' torchi di Glauco Masi, 1816).
— *Cenno Biografico del conte Sebastano de' Guidi. Introduzione dell'Omiopatia in Francia* (Napoli: Poliorama, 1837).
Romani, F., *Elogio storico di Samuello Anemanno* (Napoli: Puzziello, 1845).
Romeo, R., *Il Risorgimento in Sicilia* (Bari: Laterza, 1950).
Rossi, G., *Trattato di medicina olistica. Fondamenti metodologici biofisici e metafisici di una medicina di frontiera* (Roma: Diogene edizioni, 2012).
Sampolo, L., *In morte del professor Paolo Morello* (Palermo: Tipografia del Giornale di Sicilia, 1873).
Sbano, C., *Memorie e giudizi intorno alla poetessa Marianna Coffa in Morano di Noto* (Noto: Tipografia di F. Zammit, 1879).

Statuti fondamentali per l'Accademia sicula omiopatica di Palermo con regolamenti emanati dal magistrato accademico (Palermo: Solli, 1852).

Tebaldi, G., *Ai membri componenti l'Accademia omeopatica. Rendiconto dei lavori pubblicati dall'ab. Tripi durante il primo triennio dell'Accademia* (Palermo: G. Gaudiano, 1847).

Thuillier, J., *Mesmer o l'estasi magnetica* (Milano: Biblioteca Universale Rizzoli, 1996).

Tognotti, E., *Il mostro asiatico. Storia del colera in Italia* (Roma-Bari: Laterza, 2000).

Traetta, L., *La forza che guarisce. Franz Anton Mesmer e la storia del magnetismo animale* (Bari: Edipuglia, 2007).

Tripi, B., *Sull'andamento del cholera-morbo e suo metodo curativo nel comune di Cerba, rapporto dell'abbate Biagio Tripi* (Palermo: F. Spampinato, 1837).

— *Saggio sulla dottrina omeopatica* (Palermo: F. Spampinato, 1838).

— *Guida alla pratica omeopatica* (Palermo: F. Nocera, 1842).

— *Lessico pratico omeopatico* (Palermo: Oroteo, 1844a).

— *Statistica dei militari attaccati dal colera nell'anno 1844 e trattati omiopaticamente dal dottor B. Tripi* (Palermo: F. Lao, 1844b).

— *Appendice alla materia medica omiopatica di Jahr* (Palermo: B. Gaudiano, 1847).

— *Repertorio dei rimedi omeopatici esperimentati* (Palermo: Carini, 1852).

— *Raccolta di prose e versi scritti per le solenni esequie del Dottor Francesco Romani* (Napoli: Stabilimento tipografico, 1855).

— *Rapporto dei risultati della clinica omiopatica eretta per volere del Re, Nostro Signore, nello Spedale militare di Santa Cita in Palermo* (Palermo: F. Lao, 1856).

Vecchio. I., et al., 'Paracelso. Vita e contributo all'evoluzione delle scienze mediche, umane e neuropsichiatriche', *Acta Medica Mediterranea*, 27, 3 (2011), 153-161.

von Schoenberg, J. J. A., *Il sistema medico del dotto Ahanemann esposto alla Real Accademia delle Scienze di Napoli* (Napoli: dalla Stamperia Reale, 1822).

Mariaelena Costa

12.
A Comparison of Archive Documents: Power Conflicts Regarding the Establishment of New Bishoprics in Sicily (18th-19th Centuries)

The time frame examined is 1778-1817, a period with numerous political and social events that disrupted every institutional order in the island of Sicily, including the Church (Costa 2017: 9-34). In particular, the foundation of new episcopal circumscriptions in Sicily in the first decades of the nineteenth century presents an opportunity – from the perspective of historical research – of fundamental importance to investigate the social, political and economic interests that were enigmatic, ambiguous and conflictual.

The mere proposal for the establishment of new dioceses, put forward in Parliament in 1778 (Giarrizzo 1779: f.30), was enough to spark rivalry between several towns, create competition between various institutions and, finally, hostility between the various social classes. The story of the town of Nicosia in the Sicilian hinterland,[1] of which little is said even today, is emblematic of these conflicting dynamics; it was set up as a bishopric on March 17, 1817.

The analysis carried out here is based on the discovery of some unpublished documents found in the old documents sections of the *Biblioteca Comunale* (the municipal library) and the Diocesan Archives of Nicosia. To place the information obtained from these sources in the appropriate historical context, a historiographical survey was carried out. Thus, by combining the recovered data, a wider rereading of the event was obtained.

---

1   The city of Nicosia, which today has about 14,000 inhabitants, is located in the valley of the river Salso and nestles between the Nebrodi and the Madonie mountains. It has ancient origins and a rich history. It is adorned with many religious monuments – including the cathedral which houses a rare 14th century wooden ceiling – and numerous palaces of the barons who resided there, hence the appellation "City of the 24 barons". A particular feature of the town is the Gallo-Italic dialect that is still spoken there. Among the many books on the history of Nicosia, see: Beritelli La Via and Narbone 1852.

Specifically, the *Liber Gratiarum et Privilegiorum* is preserved in the *fondo antico* (the ancient documents section) of the Municipal Library of Nicosia; it could be defined as a sort of collection of royal and viceroy letters, public tenders, and pragmatics, thanks to requests and privileges obtained from the city, with unpublished documents dating back from the 14th to the 19th centuries.

Among the few folders contained in the Diocesan Archives of Nicosia, in the series *processo di erezione nuova sede episcopale* (establishment of the new episcopal seat), I identified an interesting unpublished report of 1808 which describes the religious and civilian situation of the city.

In order to make a broader classification of the recovered papers, it was necessary to examine the correspondence kept both at the Diocesan Archives of Messina, of which the diocese of Nicosia is suffragan, and at the Diocesan Archives of Catania, as some parts of the current Episcopal circumscription depended on it, as well as the Vatican Archive relating to the *Processus Datariae*.[2] It was also interesting to consult the section *Miscellanea Risorgimentale* of the State Archives of Catania, since Nicosia was under its jurisdiction.[3]

There was a lengthy historiographical debate regarding the conflict between the Church and the State focusing on the aspects of *Legatia Apostolica*, especially on the last years of its validity. It was of fundamental importance to consult the publications of Michele Amari (Amari 1872: 306-307), Domenico Lancia di Brolo (Lancia di Brolo 1880), Luigi Boglino (Boglino 1899), Gaetano Catalano (Catalano 1950), Edith Saurer (Saurer 1972: 33-43), and the most recent works by Salvatore Vacca (Vacca 2000). Above all, for the purposes of this research, the following works were essential: the studies of Monsignor Gaetano Zito, episcopal vicar for the culture of the diocese of Catania, author of numerous publications

---

2   These are informative processes relating to candidates for the episcopate of the Vatican state offices. The fund was organized from 1622 until 1830. In the aftermath of the Council of Trent, the Church became more and more attentive to the state of the vacant church and to promoting it, in order to have a complete picture of the pastoral government and the economic situation of the region. Nestola 2013: 586.

3   The Stewardships started in Sicily in 1817 following the administrative reform of the sovereign, in order to better control the suburbs. The ancient division of Sicily into the three valleys of Mazzara, Demone and Noto was abolished and replaced with seven stewardships: Palermo, Catania, Messina, Agrigento, Syracuse, Trapani and Caltanissetta. Nicosia was under the stewardship of Catania along with Caltagirone. See Frasca 2014: 59-107.

concerning the History of the Church of Sicily, and the work of Father Salvatore Gioco which deals with the history of the Nicosia diocese.

At the end of the eighteenth century in Sicily, the territorial organization of the Church was still linked to the Norman order, i.e. it was divided into nine dioceses,[4] all located along the coasts of the island. The sovereign in Sicily enjoyed an important privilege, obtained in the Norman period: the *Legatia Apostolica* (Apostolic Legacy), which allowed him, among other things, to manage the episcopal duties, the administrative, legal and financial aspects of the dioceses. The popes repeatedly tried to repeal or reduce these privileges but without success, and often various jurisdictional conflicts broke out between the State and the Church. The most striking occasion is certainly the one that passed to history as the "Lipari controversy"[5], which

---

[4] They were: Palermo, Messina-Troina, Catania, Syracuse, Agrigento, Mazara, Monreale, Cefalù and Patti. In 1775, on the death of Mons. Testa, Monreale was suppressed by Ferdinand III of Sicily and grouped with Palermo, so the number of dioceses was reduced to eight. The Metropolitan Archbishop of Monreale, originally from Nicosia, had distinguished himself for his culture – his research that culminated in the famous publication on the chapters of the Kingdom of Sicily remains unsurpassed – but above all for having brought the seminary of his diocese to high scientific, literary and historical-philosophical levels, so much so that Monreale was given the name of the *Athens of Sicily*. On his death, the sovereign obtained from Pius VI the amalgamation *Aeque principaliter* with Palermo, dividing with the Archbishop the annuity of the suppressed diocese and committing his portion to the preparation of the fleet; this measure provoked such strong reactions that the king was forced to go back on his decision and give Monreale its autonomy back in 1802. See Zito 2009: p. 539.

[5] The story is clearly described by G. Zito: «The name derives from the place where the episode that started the battle took place. The incident was trivial in itself, but the initial protagonists and others involved all had a strong awareness of the jurisdiction connected with their institutional role. On 22nd January 1711, in Lipari, two guards confiscated about 800 grams of chickpeas from Nicola Buzzanca at the bishop's refectory as payment for the municipal tax owed by the bishop. The bishop, the Benedictine Nicola Maria Tedeschi, presented his objections because he believed that his ecclesiastical immunity and his jurisdiction had been violated. The two guards returned the chickpeas, but the bishop demanded a public apology from the jurors of the town of Lipari. As they refused, Tedeschi issued a decree of excommunication for the guards. They appealed to the Tribunal of the Royal Monarchy. The judge, of course, declared the bishop's decree to be unjust and therefore void. Supported by other Sicilian bishops, the bishop appealed to Rome. Here it was declared that the court did not have the power to absolve the excommunications, which was reserved for the pontiff, and this ordinance was ordered to be published in all the dioceses of the island. The Roman decree, however, was obviously not granted the *regio exequatur*. From this moment, and for about twenty years, there was a succession of episcopal, pontifical and royal

began in 1711 and was resolved only in 1728 with the so-called *concordia benedettina*[6] (Benedictine concord); however, the controversy between regalists and curialists did not end with this.

The need to set up new dioceses was officially justified by the increase in population, due to the foundation of new cities, but also by the geography of the island; the inland and mountainous areas made it difficult for priests to move around to meet the pastoral needs of the people (Giarrizzo 1779).[7] In particular, as was pointed out in the Consultation of the Deputation of the Kingdom of 1778 (ADN *processo*: 1-5), the diocese of Messina presented the greatest need, as it was the largest, covering the Val Demone almost entirely.

With these premises, the sovereign was asked to create new dioceses on the occasion of the Parliament of 1778. The king agreed but did not give permission to proceed until the Deputation has given him a detailed plan of the territorial division to be made and the criteria for assigning the bishoprics (Gioco 1972: 17). The project started from the diocese of Messina because, as we have already said, it was the largest and, moreover, it was vacant (ADN *processo*: 20,1-5); what raised more protests was the proposal to establish the bishopric in the inland territory (Zito 2009: 89). The only uncertainty was the seat, a choice that initially fell equally between the cities of Troina[8] or Nicosia (ADN *processo*: 20,1-5), sparking the rivalry between the two cities that aspired to the honor of the bishopric.

Therefore, following the multiple requests and pressures he received, Ferdinando worried about losing the people's goodwill, and «thought it wise and prudent that things remained as they had been for so many centuries» (ADN *processo:* 19, 919). However, the competitive climate did not let up; in fact, for fear that one day the plan could be taken up again, the

---

interventions each defending their own prerogatives, rejecting jurisdiction from the opposing side». See Zito 2009: 79-80.

6 This is the bull *Fideli ac prudenti dispensatori*, issued in 1728, with which Pope Benedict XIII restored the privilege of the *Legazia* but obtained from the sovereign Charles VI the power to stipulate agreement pacts and to intervene in the affairs of the Church in Sicily. See Gambasin 1979: 7.

7 For example, the presence of the bishop was required to celebrate the rite of Confirmation, the sacrament necessary to hold a wedding ceremony.

8 The City of Troina, with about 9.400 inhabitants today, is located in the Nebrodi park and near the Ancipa lake. The castle was a vital outpost for Count Ruggiero to conquer the island. It is known for being the site of the bloody battle, called "The Battle of Troina", between the Allies in Sicily and the German resistance in 1943. The city was reduced to rubble and the photos of Robert Capa are reminders of that massacre. For the history of Troina see: Tramontana and Cantale 1998.

cities involved in the project always kept their guard up, sometimes raising new controversies.[9]

Indeed, this is what happened in the Parliament of March 24, 1802 (Gioco 1972: 119). On the occasion of the death of the bishop of Syracuse[10], who died during the pastoral visit to Caltagirone, the problem of the territorial vastness of the dioceses and the need to create new ones was again highlighted, in particular following the dismantling of those of Catania, Syracuse and Messina, which in the last century had been the subject of numerous *Licentiae Populandi* and therefore recorded an increase in population density (Longhitano 1979).

The project was resumed under very favorable political circumstances. Sicily had indeed made great sacrifices, responding generously to the constant requests for money made by the King for the war against the French and for the maintenance of the royal court in Palermo[11], and as was customary, together with the donations there was the presentation of the so-called *graces*, which the sovereign most often approved.

Thus, when the subject was reopened, the candidacies of the two aforementioned cities were analyzed in detail. In Troina's favor, the historical motivation was strong: the *reductio ad pristinum* of the ancient bishop's seat, founded by Count Ruggero in 1081, had never actually been canceled, so the bishop of Messina still held the title of *Archiepiscopus Messanensis et Episcopus Traynensis (ADN processo:* 20, 5-9). On the other hand, the demographic and economic conditions of Troina were not considered suitable to support the salary of the new bishop, unlike Nicosia, a town that boasted an abundance of souls, food and wealthy men (Beritelli La Via and Narbone 1852: 8). Moreover, in 1743, during the plague that raged in Messina, Nicosia was the seat of the vicar general in *partibus*

---

9 There were many cities that aspired to the episcopal seat; some had had it in the past, others desired it for prestige. In addition to Nicosia, opposed by Messina and Troina, see the case of Caltagirone opposed by Syracuse and Lentini, and Piazza Armerina whose candidacy was opposed by Catania and Castrogiovanni (today Enna). Noto, Taormina and Trapani were among the other cities that were nominated.

10 Monsignor Alagona fell ill during his pastoral visit, and since it was difficult to take him back to Syracuse he remained in Caltagirone where he died. Zito 2009: 322.

11 King Ferdinand, forced to flee Naples, found refuge in several places, including Piazza Armerina; he was hosted in the palace of the Trigona della Floresta family. The warm welcome he was given was a reason why the diocese was established there; the sagacious moves of the local dignitaries also played a part. Tandurella, *Piazza Armerina*, in Zito 2009: 698.

montanearum (ADN *processo*: 18, 305), Deacon Carlo Speciale, who operated in the mountainous area of the diocese. This occasion served to demonstrate the importance of the city in the internal territory.

The competition between the two cities also depended on who could make the abundant donation of 1200 *onze*, for the bishop's house, the seminary and the bishop's refectory. Yet only Nicosia could offer this amount immediately, since Troina was in financial difficulty so it proposed the offer in installments.

Between 1778 and 1820, Nicosia was among the most influential cities in the hinterland. Besides the demographic density (18,000 inhabitants), noblemen (209 noblemen, 24 Barons of whom 10 held the special honor "*mero e Misto Impero*", 8 had a noble fiefdom, and 6 had a title), and clergy (185 priests, 3 parishes, 3 female monasteries and 1 College of Maria, and 4 Convents), the *Universitas* managed a yearly patrimony of 35,108 *scudi*. It is also important to remember the large number of services it offered compared to neighboring towns. It was the seat of the judicial offices: civil, criminal, appellation, *ideota* and the delegate of the Royal Monarchy and Apostolic Legacy. As many as 20 notaries resided in the city. Due to the administrative reform of 1817, it became under the jurisdiction of Catania. In addition, it was head of a military district and alone had 44 Knights and 227 infantrymen.

There was a very active *Monte di Pietà* (prerequisite for the establishment of a bishop's seat), with an income of 800 *onze*, which also ran the hospital – with an income of 700 *onze*, in which 15 doctors practiced – and the town prison.

Its market, which was on the circuit of fairs in the region, was a place of commercial exchange for the whole district; moreover it had a *Colonna Frumentaria* (corn exchange) and above all a warehouse which, during the years of famine, provided grain not only for its own population but also the nearby towns.

As for urban planning, the Senate was constantly involved in road maintenance and, moreover, at a cost of 200 *onze,* a project by Interguglielmo, Artinelli and Marvuglia had created a public aqueduct, supplying two drinking troughs. Finally, following the royal reform of public studies, there was also the *la Casa degli Studi* (the House of Studies) (Raffaele 2005), with as many as six professorial chairs (Gioco 1972: 202-204).

Troina was now out of the running and began to support Messina with all its might, refusing to dismantle the diocese, especially for the economic impact that it would have on the city.

In opposition to Nicosia's candidacy, the internal rivalry between the two most important Collegiate Churches, S. Maria and S. Nicola, was highlighted. This was a diatribe that had distant roots and to which a solution had been found on July 18, 1589 with the *Lodo di Orosco*, which decreed that the functions of the mother church should be alternated (La Via 1898: 478-515). The conflict between them was rekindled because they both aspired to the title of Cathedral.

Despite these criticisms, in 1802 the final decision was in favor of Nicosia, but it had to wait until March 17, 1817 for the official institution, following the release from prison of Pius VII. It was established as a suffragan of Messina, initially comprising 19 municipalities, then reduced to 12 with the creation of other bishopric districts in 1844.

A further reason for opposing the new diocese concerned the design of regional boundaries, which obviously included changes and cuts in other dioceses, choices that «did not always follow pastoral motivations» (Zito 2009: 551).

In fact, Ferdinando saw the right opportunity to implement his reform project, which made the administrative offices coincide with those of the bishop wherever possible, following the model of the French reform that identified the provincial capital with the capital of the diocese (Zito 1995: 68).

The Monarchy tried to bind itself to the Church, as shown by the many measures of ecclesiastical politics of the Bourbons: the gradual reopening of convents and monasteries suppressed during the decade of French rule; authorizations granted to ecclesiastical bodies for the acceptance and purchase of inheritances and legacies; the imposition of the Catholic-Roman religious character on primary education; and finally, the prohibition of the publication of anti-religious books (Jemolo 1967: 126-127). The nomination of the new bishops should therefore be read in a political key and included in the reform project established by Ferdinando I (Zito 1995: 78; Gemmiti 1989; Faggioli 2001: 221-256).

In the reform of the administrative structures of the island, and therefore of the ecclesiastical ones, the bishops were set «organically in the bureaucratic apparatus of the state, affecting the care of its souls» (Gambasin 1979: 80).

In addition, the decree of April 5, 1818 explicitly renewed the prerogatives of the Court of the Monarchy of Sicily, already secretly reiterated in the same agreement. The bishops were appointed by the sovereign and they swore loyalty to him. The conclusive part of this oath is significant where the bishop declares that «if in the diocese or elsewhere, I learn that anything is to the detriment of the State, I will inform His Majesty» (Maturi 1929).

It should be noted that the sovereign's choice of the appointment of new bishops and new bureaucrats favored the social rise of the emerging bourgeoisie and the establishment of a new local patrician class, to the detriment of the old baronial class, which could thus be controlled. It is also interesting to underline how the aristocracy had well placed its followers in the leadership of the clergy, above all with the creation of new dioceses, in the hope that they could intervene. In fact, there were several conflicts between the bishops and the chaplains of the cathedrals that arose from the transformation of the pre-existing colleges, and also with the local dignitaries, who, having played an important role in the request for the establishment of the bishop's seats and in their endowment of assets, also interfered in the internal life of the Church (Pennisi 1995: 96).[12]

However, already in the eighteenth century, the Sicilian ecclesiastical world had received a strong cultural and pastoral drive thanks to a group of particularly proactive bishops with Enlightenment sympathies, linked to the Neapolitan reformist circles, such as Serafino Filangieri in Palermo, Francesco Testa, originally from Nicosia and bishop in Syracuse and Monreale, Salvatore Ventimiglia in Catania, and Gabriele Di Blasi in Messina. In the pastoral field, these bishops gave a strong impetus to evangelization and the catechetical training of the people. With the expansion of the popular missions, preached by Jesuits, Franciscans and Redemptorists, catechism was very popular in the eighteenth century; some of those same bishops who promoted it wrote the catechism text in Sicilian for the use of their own diocese. Indeed, this catechetical commitment helped to mark a reversal of the trend among the secular clergy. Despite the fact that the reform decrees of the Council of Trent had ratified the urgency to give a strictly pastoral shape to the life and activity of diocesan priests, they persisted in having a greater propensity for cultural activity and a marked lack of theological preparation that in essence left the task of preaching to the regular clergy.

This is why the Bourbon legislation became more involved in ecclesiastical matters and sanctioned a more rigid control on the life of the religious. Therefore, there were not only pastoral needs at the basis of the establishment of the new dioceses, although they have been formally highlighted; in reality, the decision of the sovereign was politically motivated (Zito 1995: 68). Consequently, if on the one hand the Church

---

12   These conflicts would result in the revolution of 1820, which by now must be reread as a defense of its privileges by the ancient Sicilian nobility. In this regard, see: Barone 1991: 175-198.

felt protected, on the other hand it felt deprived of its own autonomy. Therefore, in order to reaffirm its identity and protect its own power, in competition with the new local political authorities, a new game was to start among these forces that would lead to the transformation of urban space (Iachello 1995: 239).

Bibliographical references

Amari, M., *Storia dei Musulmani di Sicilia* (Firenze: Le Monnier, 1872), vol. III.
Archivio Diocesano di Nicosia (ADN), *Fondo Cattedrale. Processo di erezione nuova sede episcopale,* vol. 18 – 20.
Avarna di Gualtieri, C., *Ruggero Settimo nel Risorgimento siciliano,* Documents used to write the history of Sicily published by the Sicilian Society for the History of the Fatherland, Second Series, Sources of Sicilian Law, Vol. VIII. Parliamentary Act decreed on 7 November 1812 by the *Protonotaro* of the Kingdom of Sicily.
Barbato, A., *Documenti inediti da servire per la storia di Nicosia nel Medio Evo* (Nicosia: Tip. Del lavoro, 1919).
Barone, G., 'La Rivoluzione e il Mezzogiorno. Monarchia amministrativa e nuove élites borghesi', in G. Milazzo and C. Torrisi (eds), *Ripensare la Rivoluzione francese: gli echi in Sicilia* (Caltanissetta – Roma: Sciascia, 1991), pp.175-198.
Beritelli La Via, G., and A. Narbone, *Notizie storiche di Nicosia* (Palermo: Stamperia Giovanni Pedone, 1852).
Boglino, L., *La Sicilia Sacra. Pubblicazione periodica* (Palermo: Boccone del Povero, 1899), vol.1.
Calisse, C., *Storia del Parlamento in Sicilia dalla fondazione alla caduta della Monarchia* (Torino: unione Tipografico Editore, 1887).
Catalano, G., *Le ultime vicende della Legazia Apostolica di Sicilia. Dalla controversia liparitana alla legge delle Guarentigie (1711-1871)* (Catania: Facoltà di Giurisprudenza, 1950).
Costa, M., 'Fondazione della Diocesi di Nicosia', in Aa.Vv., *I Vescovi della Diocesi di Nicosia. Quattordici Pastori...duecento passi della nostra Chiesa!* (Nicosia: Edizioni Creativamente, 2017), pp. 9-34.
De Stefano, A., *Il Parlamento siciliano* (Firenze: Almannacco Italiano, 1948).

Faggioli, M., 'La disciplina di nomina dei vescovi prima e dopo il Concilio di Trento', *Società e Storia*, 92 (2001), 221-256.

Frasca, E., '"Questa sì deliziosa e cospicua parte dell'Isola". Il Valle di Catania nei discorsi degli intendenti (1819-1854)', *Annali della Facoltà di Scienze della Formazione*, 13 (2014), 59-107.

Gambasin, A., *Religiosa magnificenza e plebi in Sicilia nel XIX sec.* (Roma: Edizioni di storia e letteratura, 1979).

Gemmiti, D., *Il processo per la nomina dei vescovi. Ricerche sull'elezione dei vescovi nel sec. XVII* (Napoli-Roma: LER, 1989).

Giarrizzo, D., *Codex Siculus* (Panormi: Typ. SS. Apostolorum, 1779), Tom. I, Lib. I.

Gioco, S., *Nicosia Diocesi* (Catania: Musumeci, 1972).

Iachello, E., 'Il controllo dello spazio urbano', in G. Zito (ed.), *Chiesa e società in Sicilia – i secoli XVII-XIX* (Torino: SEI, 1995), pp. 237-254.

Jemolo, A. C., *Chiesa e Stato in Italia negli ultimi cento anni* (Torino: Einaudi, 1967).

La Via, M., *Rivalità e lotte tra Mariani e Nicoleti in Nicosia di Sicilia* (Palermo: Tip. Lo Statuto, 1898), *Archivio Storico Siciliano*, vol. XXIII, 478-515.

Lancia di Brolo, D., *Storia della Chiesa in Sicilia nei dieci primi secoli del cristianesimo* (Palermo: Stabilimento Tipografico Lao, 1880).

Longhitano, G., Ligresti, D., Raffaele, S., Grillo, M., and R. Nicotra, *Studi di demografia storica siciliana* (Catania: Società di Storia Patria per la Sicilia Orientale, 1979).

Maturi, W., *Il concordato del 1818 tra la Santa Sede e le Due Sicilie* (Firenze: Le Monier, 1929).

Nestola, P., '«An testis sciat in qua provincia sita sit civitas» Città a giudizio: fonti processuali per un approccio multifocale di storia urbana', *Mediterranea ricerche storiche*, n. 29 (2013), 586.

Pennisi, M., 'Dusmet amministratore apostolico di Caltagirone', in G. Zito (ed.), *Chiese e società in Sicilia - i secoli XVII-XIX* (Torino: SEI, 1995), pp. 97-118.

Raffaele, S., *La bottega dei saperi: politica scolastica, percorsi formativi, dinamiche sciali nel Meridione borbonico* (Acireale – Roma: Bonanno, 2005).

Saurer, E., 'Sugli ultimi anni del Tribunale della Monarchia sicula', *Rassegna storica del Risorgimento*, 59 (1) (January-March 1972), 33-43.

Tandurella, G., 'Piazza Armerina', in G. Zito, *Storia delle Chiese di Sicilia* (Città del Vaticano: LEV, 2009), pp. 695-709.

Tramontana, S., and M. C. Cantale, *Troina: problemi, vicende, fonti* (Roma: Herder, 1998).

Vacca, S. (ed.), *La Legazia Apostolica. Chiesa, potere e società in Sicilia in età medievale e moderna* (Caltanissetta – Roma: Salvatore Sciascia, 2000).

Zito, G. (ed.), *Chiese e società in Sicilia – i secoli XVII-XIX* (Torino: SEI, 1995).

— 'Dusmet e l'episcopato benedettino siciliano tra i Borboni e l'Unità', in Id., *Chiese e società in Sicilia – i secoli XVII-XIX* (Torino: SEI, 1995), pp. 59-96.

— *Storia delle Chiese di Sicilia* (Città del Vaticano: LEV, 2009).

Cosmopolitanism and Social Conflicts

Vincenzo Cicchelli, Sylvie Octobre
13.
From Globalization of Culture to Aesthetico-Cultural Cosmopolitanism

Since World War II, cultural products have become some of the most internationally circulated goods. The flow of cultural goods between nations has increased dramatically: certain products can be found virtually everywhere on the planet (such as hit pop songs, TV series, blockbuster movies, bestselling books, etc.), and thus help to develop shared cultural imaginaries as well as global awareness. How can we grasp the cultural mutations engendered by the consumption of cultural products which mix or even employ foreign cultural codes? What are the references, and which new criteria of appreciation are engaged? Does this permanent contact with the cultural products of transnational circulation modify the link with alterity, real or imagined?

We propose a specific way to analyze the effect of the globalization of culture on the way young people see the world, by examining their seemingly banal consumption for clues to understand how they envision the world, and by introducing a specific approach stemming from cosmopolitanism and the sociology of culture: the aesthetico-cultural amateurship.

13.1 *Globalization as a cultural mosaic*

The global expansion of capitalism has granted greater significance to cultural matters (Robertson and White 2007), with some scholars even going so far as to characterize contemporary capitalism as "artistic" (Lipovetsky and Serroy 2013). The globalization of culture has become more pronounced since the mid-1990s. Cultural globalization has shifted towards a plurality of influences. Today, we can assume that the globalization of culture serves a number of different functions for countries who wish their cultural hegemony to appear unchallenged, and their narratives of the world to be so convincing that consumers embrace them without any kind of critical distance whatsoever. Hollywood enters into competition with Bollywood and Nollywood in certain geographical regions; many countries such as Brazil, South Korea and Turkey have become major producers of television series; and certain cultural forms that had hitherto been purely national in their area of distribution are now widely exported (such as Northern European crime novels, Japanese manga and K-pop). On the other hand, the circulation of cultural products and services has hastened the advent of a "global mélange" (Pieterse 2009): namely, the rise of hybridization, combining universal elements with local ones and thus allowing distinct populations to identify as members of an ethno-national community while still also seeing themselves in a broader collective group. Although hybridization is a process as old as time,[1] it has become one of the main drivers of globalization, since «the pace of mixing accelerates and its scope widens in the wake of major structural changes, such as new technologies that enable new phases of intercultural contact» (Pieterse 2009: 98). In the same vein, in his studies of pop-rock music, Motti Regev (2013) has highlighted that certain international elements (rhythm, the choice of instruments) are mobilized to support other elements that are considered to be purely national (voices, melodies, themes), ultimately creating a form of rock that is interpreted as the expression of a national genius, from Israel to Argentina. By speaking to a local public and providing it with local content, the songs produced in this manner are presented as the expression of a unified national culture. And yet, these cultural products are part and parcel of the global aesthetic of the pop-rock genre, from their electric and electronic instrumentation to their sophisticated studio production techniques.

The world can therefore be understood as a cultural mosaic, both in terms of the cultural traits that are ostensibly specific to each nation, community or geographical area and the interplay of cultural comparisons that this entails, and the emphasis that is placed on cross-cutting and/or hybrid cultural characteristics.

13.2 *Capturing distance: Cultural odorlessness and cultural discount*

The globalization of culture is vigorously disparaged by the champions of cultural authenticity. Such criticisms usually follow one of two lines of reasoning. The first line of reasoning targets the cultural contents and

---

[1] The process of hybridization can be defined as 'the ways in which forms become separated from existing practices and recombined with new forms in new practices' (Rowe and Schelling 1991: 231).

their inherent qualities, arguing that the capacity for cultural contents to be appropriated by different individuals is maximized at the expense of their cultural value, resulting in what Iwabuchi calls "cultural odorlessness" (2002). Consequently, if American movies are popular, that is because they depict deculturalized and delocalized elements, and they simplify the cultural complexity of the elements that they *do* portray (Liebes and Katz 1990). In fact, sometimes cultural products are even "drained" of their cultural values to offer a variety of different meanings depending on the audience in question. It is therefore the process of deculturizing (and departicularizing) American culture that allows the country's cultural products to become global. Hollywood copies and recycles countless ideas, themes, cultural frameworks and forms of expression from other cultures – especially Asian cultures – and forges a new identity composed of transnational influences that are reflected in products that can easily be appropriated by individuals, given that cultural particularities are reduced to mere details or exotic flourishes (such as décor) that blend into an otherwise predominantly American landscape. For example, Disney revamped the story of Mulan to transform the timid and unhappy young girl of the original Chinese tale into an emancipated, Western-style heroine (Pang 2005). Frederick Wasser (1995) describes global products, whose hallmark is cultural recycling, as "rootless" and argues that fewer and fewer are truly addressed to a specific cultural community. Higbee and Lim (2010) likewise conclude that transnational products are hybrid by nature and must therefore eliminate all overly specific traits. By the same token, pop culture, like all cultures which traffic in recycling, can therefore be seen as the mark of globalized culture, which is characterized by the permanent re-appropriation of sources that are extremely varied in nature (Jenkins 2004; Mèmeteau 2014).

The second type of criticism targets the reception of cultural products, arguing that the appropriation of a given cultural content can only become possible when its cultural complexity is discounted, in turn lowering its value. Consequently, the concept of a cultural discount «suggests that foreign media have limited appeal because audiences lack the background knowledge, linguistic competence, and other forms of cultural capital to appreciate them» (Lee 2008: 118). Recent studies on the appropriation of global cultural products (Cicchelli and Octobre 2018) have nonetheless illustrated that they *can* be appropriated in a myriad of different ways depending on social conditions, and may or may not "speak" of their culture of origin, less in terms of authenticity than the dialectical distance between the "near" and the "far" (or the exotic) and one's capacity for self-reflection by means of this dialectic. As a result, an approach that focuses on global cultural contents and products does not so much deal with representativeness, or even veracity, with regard to what might be considered to be the culture of origin, but rather with the audience's ability to establish linkages between cultural contents and cultural elements such as norms, values, and behaviours. Moreover, criticisms of "poor" reception are based on the notion that, in order to achieve a high-quality reception, it is necessary to possess prior skills and knowledge. Here again, many studies have looked at the reception of cultural products and shown that significant interest can develop, even despite an objective lack of competencies for appropriation (whether cultural, linguistic, or other). Research has also demonstrated that reception always involves the co-creation of meaning: individuals do not merely receive pre-existing meanings nestled within cultural products. We cannot understand the popularity of Japanese manga and South Korean K-pop in France unless we concede that curiosity towards others can transcend the limits of pre-existing cultural proximity (ostensibly reducing the distance between the Self and the Other). The accusation of cultural discount (which resurrects the spectre of cultural legitimacy), assumes that the "banal" forms of appropriation in a globalized context are of poor quality, and ultimately re-establishes a certain distance (this time, in terms of cultural legitimacy). France is not home to particularly large or prominent Japanese or Korean communities, as is the case in a country like Brazil, for instance. Nor does its history explain any specific ties to these two countries, unlike that of the United States, for example. And yet, France is the second-largest consumer of Japanese manga and has been since the 1990s.

13.3 *Aesthetico-cultural cosmopolitanism as a specific approach to cultural globalization*

Other scholars have established a link between the globalization of culture and the emergence of so-called aesthetic cosmopolitanism, both in terms of the cultural offering and how it is received and interpreted (Papastergiadis 2012; Regev 2013). In international scientific literature, the adjective "cosmopolitan" has been used in a myriad different ways since the emergence of cosmopolitan studies as a viable theoretical approach (Delanty 2009) – and according to some, as a truly new paradigm – and an empirical field of research in the social sciences. While we shall not attempt to provide a comprehensive overview of the many different applications of the word (Vertovec and Cohen 2002; Skrbis and Woodward 2013), we shall focus on trying to understand how individuals who are defined as "cosmopolitans" develop a relationship to the world in our contemporary societies characterized by globalization. Cosmopolitanism must now be seen as a concrete lifestyle that countless individuals can achieve through a plethora of points of contact with cultural difference, offered by global capital flows, as well as the circulation of norms, aspirations, imaginaries, products, services (especially in the media and cultural sphere) and persons. The fact that digital cultural industries are more global than ever means that individuals are no longer just consumers but also agents of cultural globalization, since they participate in the production and diffusion of increasing volumes of cultural content, from tutorials to self-published and collaborative creations (Octobre 2018). In addition, the growing use of cultural resources to define one's identity signifies that aesthetico-cultural cosmopolitanism has become one of the most easily available and thus banal forms of cosmopolitanism, as illustrated by Mica Nava (2007) in her study of London women and fashion at the beginning of the 20th century. Finally, by marketing cultural difference, contemporary consumerism identifies exoticism as an engine of desire for further consumption (Holt 1998; Emontspool and Woodward 2018).

Any analysis of cultural globalization requires a specific approach, one that takes into account what individuals *do* with globalized cultural products. What we have termed aesthetico-cultural cosmopolitanism is an orientation towards others that presupposes a certain degree of familiarity with aesthetic standards and cultural codes that fall outside one's immediate sphere of belonging, a familiarity which can be developed through the consumption of globalized cultural products and contents (Cicchelli and Octobre 2018). At the individual level, aesthetico-cultural cosmopolitanism is

> a cultural disposition involving an intellectual and aesthetic stance of "openness" towards peoples, places and experiences from different cultures, especially those from different "nations" (Szerszynski and Urry 2002: 468).

This form of socialization to otherness through the consumption of global cultural products has both aesthetic and cultural facets, and helps individuals to elaborate imaginaries in a globalized context where cultural industries have been dominating free time and leisure activities (Baudrillard 1968), especially since the advent of the digital age.

We assume that aesthetico-cultural cosmopolitanism comprises three major components: a) an aesthetical and cultural stance of openness and curiosity to the world through consumption and imaginaries; b) a set of individual resources that allows people to encounter the world (travel, sociabilities – real and on the Internet – or linguistic skills) and aspirations (the desire to); and c) a scale of belonging to the world. Consequently, we have analyzed cosmopolitan amateurship as an aptitude for inhabiting the world – e.g. to feel, reason about it and negotiate it through aesthetic-cultural cosmopolitanism – grabbing all the opportunities possible in the real or virtual world to meet and seize alterity. A self-assertion developed to inhabit the world thanks to aesthetic-cultural cosmopolitanism is the result of the dynamics of these three dimensions.

13.4 *An innovative approach*

The originality of the research lies in the combining of quantitative and qualitative data; the former enable us to describe the sociological morphology of aesthetico-cultural cosmopolitanism, whereas the latter give access to young people's narratives about their use of cultural contents to relate to the world. The research comprised:

a) an in-person survey conducted in France among a representative sample of young adults aged 18 to 29 years old living in France (N = 1,605, stratified by age, sex and size of urban unit), specially designed to understand how young people appropriate internationally disseminated cultural products for themselves;

Aesthetico-cultural cosmopolitanism with regard to consumption patterns is hence defined by the intersection of two dimensions:

i) The morphology of cultural repertoires, described using three kinds of indicators:

- the proportion of products according to national origin (four options: exclusively or primarily French; equally French and foreign; primarily or exclusively foreign; not applicable);
- preferences for the language of origin (six options: exclusively in French; primarily in French; equally in French and in a foreign language; primarily in a foreign language; exclusively in a foreign language; not applicable);
- preferences for products according to their national origin (three options: preference for French products, preference for foreign products, no preference).

ii) The morphology of transnational imaginaries in terms of artists and monuments, also described using three types of indicators:

- degree of familiarity: young individuals were asked if they were familiar with each item on each list (directly or indirectly) (yes = 1; no = 0);
- taste: individuals were then asked if they liked each item or not (yes = 1; no = 0);
- associated value registers: individuals were asked what value they attributed to artists and monuments (from 0 to 5), invoking four different value registers associated with global imaginaries: an aesthetic register (beauty, genius); a cultural register (defining one's vision of the world); a national register (representative of a country); or a universal register (belonging to human heritage).

b) 43 in-depth interviews with young people regarding their cultural patterns, their global interests, and their relationship to the world.

We were thus able to simultaneously measure the degree of openness of cultural repertoires and imaginaries and their strength in terms of appreciation/rejection, by looking at the choices made with regard to forms of consumption and the aesthetic value awarded according to varying registers.

By doing so, we addressed 4 streams of questions:

1) Combining consumption and imaginaries: our approach goes beyond merely analyzing the internationalization of cultural repertoires or the mobilization of cultural competencies. It analyses both the individual and social morphologies of aesthetico-cultural cosmopolitanism and the contribution of cultural consumption to forging imaginaries of the world.
2) Cosmopolitanism on the ground: using an original methodological protocol, we identify traces of the great abstract principles of cosmopolitanism in the little things of everyday life (Skrbis and Woodward 2013), in the daily lived experiences of social actors, in their banal behaviours (DeNora 2014), which are so many examples of "ordinary cosmopolitanism" (Lamont and Aksartova 2002; Kendall and others 2009).
3) Configurations instead of ideal types: Cosmopolitanism is mostly discussed in terms of ideal types, using dichotomies such as aristocratic/popular, open/closed, archetypal/ordinary, creative/industrial, diversity/homogenization, and so on. However, we have chosen a different

approach that seeks to analyze a continuum of "impure forms" (Beck 2011) of cosmopolitanism instead of searching for "pure" models that would serve as a standard of measurement.
4) The role of social stratification and individual skills. For some, cosmopolitanism reflects the ideology of an elite (Calhoun 2003) and seems to embody a discredited Eurocentric and liberal ideology in a newly-dangerous guise. The cosmopolitan figure is often negatively identified with the privileged mobile elite, whose cultural curiosity reflects a lack of loyalty to any community, as well as only superficial concern for humanity and global issues. Much like cultural sociology has often overlooked the impact of the globalization of cultural products and cultural imaginaries, the sociology of aesthetic cosmopolitanism has often neglected to identify the social determinants of openness.

Our approach seeks to address these two missing elements by analyzing the spectrum of aesthetico-cultural cosmopolitan among French youth along a socially stratified continuum.

13.5 *A widespread new "good taste" among young people*

The quantitative data analysis highlighted the existence of 5 configurations of cosmopolitanism, each of which corresponds to a different response to cultural globalization: the first is immersion without any particular intentionality (inadvertent cosmopolitanism, 34%); the second is engagement, which can take two forms (specific cosmopolitanism, which occurs primarily through reading, 32%; and cosmopolitan fandom, which is marked by a general stance of openness to diversity, 17%); the third response is rejection (national preference, 11%); and the fourth is retreat (impossible cosmopolitanism, 6%). These results indicate that cosmopolitanism is a major generational phenomenon in how young people relate to culture in a global world.

Even though aesthetico-cultural cosmopolitanism is widespread among young people, the issue of social status, age and gender cut across all five configurations, highlighting their internal variations. Hence, *inadvertent cosmopolitans* tend to live in rural areas or small/medium-sized cities and have low educational levels. The majority of them are in the workforce, where they have low-skilled and low-paying jobs such as blue-collar workers or office employees. They are more likely to be married and saddled with family duties (moreover, some of these individuals are single parents). They come from uneducated, low-income backgrounds where traditional family structures, including stay-at-home mothers, are the most prevalent, which naturally has repercussions on how they view work and leisure. The intergenerational social reproduction of family structures is generally supported by their choice of partner, who is likely to be on a similar low education level and quite familiar with the world of work and its difficulties (a high percentage of their partners have been unemployed at some point). Inadvertent aesthetic cosmopolitans, who are largely fluent in French thanks to their family environment, include young individuals from a Francophone linguistic tradition as well as the children of immigrants from Africa.[2] Thanks to the school curriculum, they benefit from learning two foreign languages, although the majority of them state that they have not fully mastered those languages.

The *specific cosmopolitans* group, which skews slightly female, includes more individuals from the 25–29 age range with some of the highest education levels in the sample, and comprises the largest number of senior managers, executives and intellectual professionals. Endowed with both academic and economic resources, individuals in this group are also more advanced in their life trajectories (more than two-fifths already have children). These individuals are on track to reproduce their social situation: their parents are similarly educated and belong to the middle or higher socio-professional classes. In addition, their marital homogamy means that they chose partners with similar, or slightly lower status socio-economic profiles (white-collar employees, for instance). They live primarily in large cities, where academic, cultural and professional resources are concentrated. This group contains the largest number of individuals who master foreign languages: while a little more than a quarter of this group only speak French, more than one-third have mastered two foreign languages, and a slightly fewer than another third are fluent in three foreign languages. This multilingualism is largely the product of education, since most of these individuals come from French-speaking households and families. It should be noted, however, that for specific cosmopolitans, education produces a higher level of multilingualism than for the previous group.

The members of the *cosmopolitan fans* group, which skews slightly male, are the youngest in the study, those with the highest education levels and those most likely to live in an urban environment. Most of these young individuals are still students, but those who work are more likely than the

---

2 Second or even third generation, since 90% of the young people in this group were born in metropolitan France and 2% in overseas territories.

average to have entered into higher professional positions. Their mothers and fathers are also among the most educated in the sample, the majority of them working in higher socio-professional occupations and paired in highly homogamous couples. This homogamy is replicated in the choices made by the young individuals who have a partner. The young people on the edges of the cosmopolitan fans group largely grew up in a French-speaking environment, but to a lesser extent than the previous group.[3] Their multilingualism is also thanks to education. This group includes the largest number of individuals who state that they are fluent in three languages, as well as the largest number of those who state that they learned three languages in school or in their professional lives.

The individuals in the *national fans* group, which skews slightly female, are rather less educated. When they are employed, they are generally manual workers and other blue-collar workers. Close to one out of every seven has been unemployed. With fewer academic qualifications and thus encountering more difficulties in the labor market, close to half of these individuals also have families (and some are single parents). The living conditions experienced by those in this group are thus more difficult than those for the other groups; as a result, their relationship to leisure activities is different. These individuals come from less educated families who are steeped in working-class (largely male) culture, where stay-at-home mothers are the norm and if they work, women are usually in intermediate occupations. Here again, social reproduction goes hand-in-hand with homogamy: those who are in couples have chosen partners with a similar low education level and who belong to the same socio-professional classes (primarily the working class). These individuals can be found in urban areas of all sizes. This configuration displays poor language skills, including with regard to French. While young people in this group might initially benefit from linguistic environments that are richer than the average on account of their families (in particular because of the presence of Arabic), they benefit slightly less than others from the two-language curriculum at school, in part because their schooling tends to have been shorter (many stop before the baccalaureate). Most importantly, these individuals are the most likely to state that they only speak French, which limits their modes of appropriation of cultural products in foreign languages.

Finally, the *impossible cosmopolitans* group, which skews heavily male, includes rather well-educated individuals who stand out because they come from heterogamous families (where their parents have highly

---

3 "Only" 86% of these individuals were born in France.

different qualification levels and modes of labor market insertion). As a result, these young individuals are confronted with hybrid socio-cultural models and work ethics, linked with specific insertion trajectories: some of them will become blue-collar workers, but most will become tradespersons and small business owners, jobs for which the link between educational level and income level is in fact the least linear. The partners chosen by these individuals also reproduce their parents' heterogamy, as many are married to senior executives and managers. Members of this group primarily live in large cities. These individuals have significant language skills: close to six out of ten state that they are fluent in several languages (one-third are fluent in two and one-fifth in three). These young individuals are more likely than average to come from working-class immigrant families (5% were born in Northern Africa, 2% in Francophone African countries, 2% in other European Union countries and 1% in other African countries). They are also the most likely to experience multilingualism within their couple, even if the majority still speak French with their spouses. However, these language skills are not mobilized in their forms of cultural appropriation.

Our analysis thus leads us to reject the idea that cosmopolitanism is a rare phenomenon, the exclusive purview of a small number of highly educated, privileged urban individuals – an argument that has already been put forth in a number of qualitative studies that revealed the existence of different forms of openness to diversity, including among the lower classes (Lamont and Aksartova 2002): a humanist mode, which is primarily intellectual; a populist mode, which is relatively lowbrow and linked to eclectic but indiscriminate tastes; a practical mode, that stems from curiosity and develops through personal experience; and, finally, an indifferent mode, which combines a little bit of each category (Ollivier 2008).

The spread of aesthetico-cultural cosmopolitanism can be explained in two different ways: firstly, by looking at the structural transformation of the world that young people live in, which has become more urban, more educated, more mobile and more multicultural; and secondly, by examining the transformation of the cultural universe for young people, which is now defined by eclectic tastes (omnivorism) (Octobre 2014).

Therefore, we can see two groups of people with very different experiences of youth. In one of them, adolescence ends when one enters the labor market and marries, relatively young, thus putting an end to any experimentation (if such had even taken place). In another experience, youth is stretched out as individuals follow longer and longer academic paths, at least partially with a view to "discovering oneself", a widespread ambition among the middle and upper classes. The stratification linked to education levels and lifestyles is combined with the various forms of stratification tied to the cultural offer available where one lives: individuals living in rural areas or small towns are more likely to be inadvertent cosmopolitans, whereas the inhabitants of large cities, where the cultural offer is myriad, tend to be more active cosmopolitans (specific and fans). More than one-third of inadvertent cosmopolitans reside in small cities, whereas more than two-fifths of cosmopolitan fans live in large cities; a quarter live in Paris and the surrounding region.

13.6 *A new figure of amateurship*

The narratives help us understand the way young people use cultural contents to relate to the world. The role of the media-culture (Macé and Maigret 2005) in the development of this aesthetico-cultural cosmopolitan stance is central for two main reasons: first because cultural media industries reformulate cultural legitimacy, in a post-national consumption context, especially among young people, as evidenced by the emergence of their taste for foreign television series (Glévarec 2012); and second, because digital technology and the Internet have emphasized the various points of connection between groups and cultures, as well as multiplying the impact of the media on individual and collective imaginaries, on a global scale (Appadurai 1996). It is therefore unsurprising that young people most often refer to works of fiction, as they serve a dual purpose:

> they provide us with experiences that we have not had, and shed light on those that we already know. [Fiction] warns us, informs us and educates us. Sometimes, it even comforts us (Winckler 2012: 98).

Individuals can independently develop modes of reception that go beyond what producers and programmers had intended (Lessig 2008).

The narratives shed light on the skills and competencies developed by mediated encounters with cultural alterity and their condition of emergence. They are made possible thanks to one key mechanism – aesthetico-cultural *mise en genre,* or the categorization of cultural products according to their national origins and their attributed aesthetic characteristics (Cicchelli and Octobre 2018). As a result, individuals do not escape a kind of "methodological nationalism" in terms of cultural reception, even when the characteristics seen as "national" are in fact hybrid in nature and do not adhere to strictly ethno-national cultural forms (as might be classified by an anthropologist for example).

The process of *mise en genre* is accompanied by the identification of what is near and far, a kind of compass to orient oneself in a highly culturalized and aestheticized world, using a certain back-and-forth between direct and mediated contacts; the former are used to verify the latter, while the latter often help to shape the imaginaries that motivate the former. In both cases, curiosity is the central motor: it feeds into desires to move beyond one's immediate circle, to be surprised, to experience wonder, discovery, and novelty. A process of self-reflection, curiosity and the desires that stem from it deliver the narrative resources that individuals need to establish their sense of belonging to a community or communities (be these ethno-national, ethno-racial, gender, social, generational or age-related communities or even just loose affiliations based around shared taste preferences). It thus becomes crucially important to know how to establish the "right amount of distance" to maintain the spark of curiosity and make moving away from one's centre of reference (decentring oneself) possible (Cicchelli 2018). Ultimately, such adjustments are the responsibility of the individual: while an individual's general social conditions (his or her economic, social, cultural and linguistic capital, as well as potential for mobility) can inform some personal stances, other attitudes are more explicitly defined by an individual's resources (biography, desires, personal aspirations, etc) (Cicchelli and Octobre 2017).

A second mechanism then comes into play which, instead of localizing the content of cultural products or the experiences encountered while travelling, serves to universalize the former. The process of universalization can take two different paths: either the path of cultural de-particularization, when a cultural product is treated as universal (works are then seen as belonging to a shared human heritage, part of the universal canon of beauty and meaning, which is implicitly founded on a certain hierarchy of tastes and knowledge); or the path of cultural re-particularization (for example, the exposition of primitive and indigenous arts or ethnographic museums). Such art forms represent a cultural and aesthetic enigma, based on the presumption of aesthetic and cultural relativism, and attest to the diversity of human production. As a result, these skills allow individuals to elaborate narratives about the world which continuously establish linkages between the particular and the universal.

Ultimately, many different skills can be developed in this way: the ability to establish connections between information, perception and emotions; the ability to compare different aesthetic values and cultural norms; and the ability to suspend judgement, to develop interpretations and to express feelings, and so on. These attitudes cannot be reduced to the mere acquisition and application of "global skills", which are the purview of the globalized elite, who wield them to obtain profitable social positions for themselves.

This gives rise to a new figure of amateurship: the aesthetico-cultural cosmopolitan amateur, that refers to cultural consumption in a global world. This figure differs from previous forms of amateurism, insofar as he or she puts forward the resources provided by cultural contents to connect themselves to the world rather than the intrinsic qualities of the artworks. Hence, it implies a specific yet fleeting form of engagement in cultural contexts which are not exclusively national, and which are not strictly defined by national education systems. The cosmopolitan amateur thus appears as a central figure in youth cultural consumption in a global world (Cicchelli and Octobre 2018).

Conclusion: towards a cosmopolitan education

Recent generations are becoming increasingly cosmopolitan, as the cultural products that they enjoy come from increasingly varied backgrounds and as they hybridize their imaginaries, even if certain production centres preserve their dominance. The number of young people whose "standard of good taste" includes openness to diversity is telling: international cultural products represent the first aspect of globalization with which they have contact, as well as the first resource that they can mobilize to make sense of globalization (Cicchelli and Octobre 2017).

Therefore, aesthetico-cultural cosmopolitanism can be seen as a new sensibility in the history of ideas, a history marked successively by courtly love (De Rougemont 2001 (1972)), childhood (Ariès 1960), human rights (Hunt 2007), equality (de Tocqueville 1981 (1835–1840)), feminism (de Beauvoir 1949), the question of identity and authenticity (Taylor 1989), and so on. From this point of view, it cannot be emphasized enough that this form of neo-cosmopolitanism represents the cornerstone of contemporary civic education, much like the role played by emancipation during the Renaissance (Garin 2003), by reason during the Enlightenment (Hazard 1961; Dupront 1996), by sentiment during the Romantic period (Ciseri 2004) and by exoticism during the first half of the 20th century (Said 1979). The culture of ordinary men and women at the beginning of the 21st century would thus be necessarily cosmopolitan.

Although the effects of the globalization of cultural industries are therefore highly significant, they largely fall outside the control of educational institutions. Recent generations are increasingly cosmopolitan, as the cultural products that they enjoy come from increasingly varied backgrounds, even if certain production centres preserve their dominance.

Let us therefore consider the issue of cosmopolitan education. It is a unique phenomenon, as its primary actors are not governed by any kind of coherent, institutional curriculum. In fact, cultural industries and the media compete with each other; no overall vision of a global world or programme of universal ethics and politics can be gleaned from their actions. It is left to the discretion of young people, who through successive additions and connections, fashion contradictory opinions on sometimes identical contents, to elaborate a worldview from the existing cultural offer and their consumption of it.

What is at stake is not just the development of one's aesthetic gaze or creative ambitions, but the possibility, for all of us, to (re)build the ethical foundations of a shared humanity. This is a matter of urgency. The breakdown of the European project (of which Brexit is just the latest illustration) and growing tensions between both cultural and religious communities demand that we consider cosmopolitanism as a horizon of expectations, as a potential means to transform the social fabric, instead of merely viewing it as a fact (whether it exists or not) or a tool (in the service of political or ethical goals). This transformation of the social fabric through the inclusion of otherness via consumption has been relegated to a matter of *agency* (and which often takes the form of *Bildung* or *empowerment*) and remains sorely lacking in support from our institutions (Cicchelli and Octobre 2018).

Bibliographical references

Appadurai, A., *Modernity at Large: Cultural Dimensions of Globalization* (Minneapolis: University of Minneapolis, 1996).
Ariès, P., *L'enfant et la vie familiale sous l'Ancien Régime* (Paris: Plon, 1960).
Baudrillard, J., *Le système des objets: la consommation des signes* (Paris: Gallimard, 1968).
Beck, U., 'Cosmopolitanism as Imagined Communities of Global Risk', *American Behavioral Scientist,* 55 (10) (2011), 1346 – 1361.

Calhoun, C., '"Belonging" in the Cosmopolitan Imaginary', *Ethnicities*, 3 (4) (2003), 531-568.
Cicchelli, V., *Plural and Shared. A Sociology of a Cosmopolitan World* (Leiden/Boston: Brill, 2018).
Cicchelli, V., and S. Octobre, 'Aesthetico-Cultural Cosmopolitanism Among French Young People: Beyond Social Stratification. The Role of Aspirations and Competences', *Cultural Sociology*, 11 (4) (2017), 416-437.
— *Aesthetico-Cultural Cosmopolitanism among French Youth. The Taste of the World* (London: Palgrave, 2018).
Ciseri, I., *Romanticism 1780-1860: The Birth of a New Sensibility* (New York: Barnes ad Noble Books, 2004).
de Beauvoir, S., *Le Deuxième sexe* (Paris: Gallimard, 1949).
de Tocqueville, A., *De la démocratie en Amérique* (Paris: Flammarion, 1981 [1835-1840]).
de Rougemont, D., *L'amour et l'Occident* (Paris: UGE, coll. 10/18, 2001 [1972]).
Delanty, G., *The Cosmopolitan Imagination: The Renewal of Critical Social Theory* (Cambridge: Cambridge University Press, 2009).
DeNora, T., *Making Sense of Reality: Culture and Perception in Everyday Life* (London: Sage, 2014).
Dupront, A., *Qu'est-ce que les Lumières?* (Paris: Gallimard, 1996).
Emontspool, J., and I. Woodward, *Cosmopolitanism, Markets and Consumption: A Critical Global Perspective* (London: Palgrave Macmillan, 2018).
Garin, E., *L'éducation de l'homme moderne 1400-1600* (Paris: Hachette, 2003).
Glévarec, H., *La sériephilie. Sociologie d'un attachement culturel* (Paris: Ellipses, 2012).
Hazard, P., *La crise de la conscience européenne* (Paris: Fayard, 1961).
Higbee, W., and S. H. Lim, 'Concepts of transnational cinema; towards a critical transnationalism in film studies', *Transnational cinemas*, 1 (1) (2010), 7-21.
Holt, D., 'Does Cultural Consumption Structure American Consumption?', *Journal of Consumer Research*, 25 (1) (1998), 1-25.
Hunt, L., *Inventing Human Rights: A History* (New York: W. W. Norton & Company, 2007).

Iwabuchi, K., 'From Western gaze to global gaze: Japanese cultural presence in Asia', in D. Crane, N. Kawashima and K. Kawasaki (eds), *Global Culture: Media, Arts, Policy and Globalization* (New York: Routledge, 2002), pp. 256-273.

Jenkins, H., 'Pop Cosmopolitanism. Mapping Cultural Flows in an Age of Media Convergence', in M. M. Suarez-Orozco and D. B. Qin-Hilliard, *Globalization, Culture and Education in the New Millennium* (Berkeley: University of California Press, 2004), pp. 114-140.

Kendall, G., Woodward, I., and Z. Skrbis, *The Sociology of Cosmopolitanism: Globalization, Identity, Culture and Government* (New York: Palgrave Macmillan, 2009).

Lamont, M., and S. Aksartova, 'Ordinary Cosmopolitanisms. Strategies for Bridging Racial Boundaries Among Working-Class Men', *Theory Culture Society*, 19 (1) (2002), 1-25.

Lee, F. L. F., 'Hollywood movies in East Asia: examining cultural discount and performance predictability at the box office', *Asian Journal of Communication*, 18 (2) (2008), 117-136.

Lessig, L., *Remix* (New York: Penguin Press, 2008).

Liebes, T., and E. Katz, *The Export of Meaning: Cross-cultural Readings of Dallas* (New York: Oxford University Press, 1990).

Lipovetsky, G., and J. Serroy, *L'esthétisation du monde. Vivre à l'âge du capitalisme artiste* (Paris: Gallimard, 2013).

Maigret, É., and É. Macé (dir.), *Penser les médiacultures. Nouvelles pratiques et nouvelles approches de la représentation du monde* (Paris: Armand Colin, 2005).

Mèmeteau, R., *Pop culture: réflexions sur les industries du rêve et l'invention des identités* (Paris: La Découverte, 2014).

Mica, N., *Visceral cosmopolitanism. Gender, Culture and the Normalization of Difference* (Oxford: Berg, 2007).

Octobre, S., *Deux pouces et des neurones - Les cultures juvéniles à l'ère numérique* (Paris: La Documentation française, 2014).

— *Les technocultures juvéniles: du culturel au politique* (Paris: L'Harmattan, 2018).

Ollivier, M., 'Modes of Openness to Cultural Diversity: Humanist, Populist, Practical and Indifferent', *Poetics*, 36 (2-3) (2008), 120-147.

Papastergiadis, N., *Cosmopolitanism and Culture* (Cambridge: Polity Press, 2012).

Pang, L., 'Copying Kill Bill', *Social Text*, 23 (2) (2005), 13-153.

Pieterse, J. N., *Globalization and Culture: Global Mélange* (New York: Rowman & Littlefield, 2009).
Regev, M., *Pop-Rock Music: Aesthetic Cosmopolitanism in Late Modernity* (Cambridge: Polity Press, 2013).
Robertson, R., and K. A. White, 'What is Globalization?', in G. Ritzer, *The Blackwell Companion to Globalization* (London: Blackwell Publishing, 2007) pp. 54-66.
Rowe, W., and V. Schelling, *Memory and Modernity: Popular Culture in Latin America* (London and New York: Verso, 1991).
Said, E. W., *Orientalism* (New York: Vintage, 1979).
Skrbis, Z., and I. Woodward, *Cosmopolitanism. Uses of the Idea* (London: Sage, 2013).
Szerszynski, B., and Y. Urry, 'Cultures of Cosmopolitanism', *The Sociological Review*, 50 (4) (2002), 461-481.
Taylor, C., *Sources of the Self. The Making of Modern Identity* (Cambridge: Harvard University Press, 1989).
Vertovec, S., and R. Cohen, 'Introduction', in S. Vertovec and R. Cohen (eds), *Conceiving Cosmopolitanism: Theory, Context and Practice* (Oxford: Oxford University Press, 2002), pp. 1-22.
Wasser, F., 'Is Hollywood America? The trans-nationalization of the American film industry', *Critical Studies in Mass Communication*, 12 (4) (1995), 423-437.
Winckler, M., *Petit éloge des séries télé* (Paris: Gallimard, 2012).

Augusto Gamuzza

14.
Cosmopolitan Solidarity Practices and Identity Conflicts. The Case of International Volunteers for Development in Tanzania and Madagascar

Introduction

Over the past two decades, the concept of cosmopolitanism has received progressive scientific interest from sociology, in addition to the traditional fields of study working on the concept such as philosophy, anthropology, and political science, generating a geometric progression of studies and research that refer to this concept by identifying its dimensions, suitable methodological approaches, and functioning mechanisms within the social aggregate[1]. However, this tendency seems to be ironically (and tragically) counterbalanced by the progressive erosion of a 'romantically irreversible' idea of the globalization process – which the recent economic recession and turbulent reactions as a consequence of the SARS-Cov-2 pandemic have seriously thrown into crisis – as a perspective through which to observe contemporary social change. As highlighted by Turner (2007), modern societies are at the same time placed in globalizing centrifuge forces and new centripetal forms of isolationism; examples can easily be found in the main geopolitical big players – from Xi Jinping's China to Donald Trump in the United States, passing through Vladimir Putin in Russia, Jair Bolsonaro in Brazil and Recep Tayyip Erdoğan in Turkey – that support (and feed) a process of repolarization of the horizon of sense of our societies which activates – translating the proposal of Robert Putnam (2007) – a global hunkering-down process towards national communities. These contrasting phenomena pose new questions for sociology (and all the social sciences), in the effort to convincingly explain these two (apparently) opposite forces, renewing the interest towards a re-definition of the heuristic approach to the cosmopolitan view of daily life. Following these stimuli, sociology working in this 'contradictory' global field needs alternative and renewed frameworks to understand global

---

1   See Skrbis and others 2004; Beck and Grande 2007; Chernilo 2012; Pendenza 2017; Cicchelli 2018; Delanty 2019.

solidarity links across cultural, political, and social realms. As highlighted by Ulrich Beck in his seminal works on this issue (2000; 2002; 2006), the study of globalisation and cosmopolitanism represents a revolution in the social sciences, providing a new understanding of the "flat world"[2] but, at the same time, of its limits. Even if Beck's arguments were explicitly criticized (Skrbis and Woodward 2007) especially from the point of view of post-colonial studies asking for a "provincialized cosmopolitanism" that must be «sensitive to the voice of non-Western others and flattens the hierarchies of knowledge implicit in *global cosmopolitanism*» (Bhambra 2010: 34), it is undoubtedly true that the German sociologist highlights the fact that the cosmopolitan *weltashauung* is an inescapable necessity to critically engage with globalization and to go beyond the boundaries of state-centred disciplinary approaches typical of sociology and political science.

In other words, the present global disposition to solidarity seems to underline the *inescapable necessity* to be emphatic *only* with those who are similar to us. All the other options are conflict-oriented; the difference between who is *inside* and who is *outside* is becoming a real socio-political descriptor of our societies.

This approach to the discourse about global solidarities in present times is 'challenged' by individuals (and organizations) who think that

> the heart of our democracies lies primarily in our capacity to feel, think and act for distant others. And against those who reserve our capacity to connect to the suffering members of our nation, I contend that, insofar as Western democracies place the wellbeing of the 'human' at the centre of their political legitimacy, their communities of solidarity extend beyond the nation and encompass the world (Chouliaraki 2016)

In this sense, these people seem to show an opposite approach that understands, recognizes and underlines the otherness of the nation, its borders and its social rules, revealing the relevance of a spirit of cosmopolitan solidarity with regard to global issues and social change (Beck 2013: 8). Extending this theoretical proposal, when this form of solidarity is translated into biographies and historic/personal trajectories, what kind of outcome does it imply for the subject identity?

---

2   This expression was used by Sergio Marchionne, former CEO of FIAT and the great *deus ex machina* of the fusion between Fiat and Chrysler Automobiles during the press conference presentation of the FCA Group in 2009.

This work reports the first evidences, based on the in-depth analysis of ten life histories, conducted within the work in progress research project "Volunteering as a professional" (VolPro henceforth)[3]. The project aims to investigate the phenomenology of cosmopolitan socialisation (Cicchelli 2018) regarding solidarity by observing a particular typology of subjects – the volunteers for development in the African continent operating through NGOs – focusing in an innovative way on the dynamics that translate this experience into elements of biographical significances, highlighting contradictory elements in social identity formation. More specifically, the VolPro project aims to investigate how cosmopolitan solidarity practices – translated in the biographies of volunteers for development in Tanzania and Madagascar – impact on the processes of construction of the volunteers' social identities. The innovative and interdisciplinary nature of the project lies in two orders of reasons. The VolPro project will not limit itself to exploring the motivations that led to the experience of volunteering in Africa, but VolPro explores the progressive professionalization as a tangible consequence on social identities (and the related conflicts) that this pendular life entails when returning to the contexts of origin (in our case in Italy). The main theoretical and methodological innovations of this work can be found in: the progression of the theory on socialization and cosmopolitan solidarity dynamics (Cicchelli 2018); the stimulus to social policies aiming to develop a shareable and scalable methodological model in the world of decentralized cooperation in order to better fit the professionalization paths of NGO volunteers for development; and proposing an innovative research path to understand and manage the unintended effects that connote the human factor of development cooperation (Leenstra 2018). The project explores how the idiosyncrasies of the individual agency affect the achievement of the results of development cooperation programs, the professionalization trajectories of the subject involved in these practices, and the unintended effects of the cooperation actions.

Thanks to the collected life histories, the biographical method puts the researcher on the side of the subject, performing an important heuristic fundamental function: the researcher is able to understand the individual regarding both his/her dependence on the context and, at the same time, his/her perspective of justification of the context. In reality, «it is

---

[3] The project *"Professione Volontario"* (Volunteering as a profession) started in 2018 as research project of the *OfficinaSocialeCOPE* research laboratory (a research division of COPE ONG). For more info www.cope.it

preferably to speak of a [biographical] approach rather than a method because of the amplitude of the thematic, theoretical and methodological references to which it refers» (Melucci 2000). This approach gives an inescapable added value for sociology, enabling the understanding of the *Weltanschauung* of the social unit by exploring the diachronic experience of the individual.

14.1 *The Epistemological proposal: from cosmopolitan solidarity to solidarity of the ethic cosmopolitans*

As outlined before, the first step of the project deals with the delimitation of the theoretical field that constitutes the research background. In order to do this, it is important to create a set of research questions to orient the investigation:

- Does the concept of solidarity of cosmopolitans exist? How can we observe the concrete forms/practices of cosmopolitan solidarity? (Ontological questions)
- If it is possible to observe these practices, what are the relevant dimensions (area of investigation) in analysing this kind of phenomena? (Epistemological question)
- What are the most effective research tools in order to scientifically observe, describe and analyse this phenomenology? (Methodological question)

I tried to answer the first two questions above but the third question requires some necessary specifications in order to explain the methodological choices. As Cipolla (1993) suggests, the focus of sociological research that uses the biographical approach consists of "going around" the subject and weighing up her/his consistency and truthfulness: reawakening interest in the unfathomability of human stories. On the other hand, Cipolla reiterates that the heuristic dimension of the subject must serve as a launching pad to go further and enhance the relationship, subjectively mediated, between narrative and reality, between "lived" and "objectified" experience constructed culturally and socially.

As proposed by Cicchelli (2018) there are four possible elementary forms of what he calls *the cosmopolitan spirit* that define the approach of the individual to the globalised social reality: cosmo-esthetic, cosmo-cultural, cosmo-ethic and cosmo-political. If we compare these dimensions with one another, we can see the dialectic relationship between particularism and individualism that can be found at the basis of the cosmopolitan approach: «there are four figures of the cosmopolitan spirit related to the learning [of being cosmopolitan], the modality of approach to otherness and the enclosing of the self in a common humanity» (2018: 244 *my translation*). The cosmopolitan spirit is placed at the core of the cosmopolitan socialization that we may define as a long reflexive process through which individuals nurture and develop their relationship with otherness. This process is not a simple direct acquisition of a disposition or the activation of an individual property but, in the author's words, this process can be "contradictory and incoherent" (2018: 231). In particular, for our heuristic purposes, we will concentrate our attention on the cosmo-ethic dimension of the cosmopolitan spirit, looking for its operative translation. In this sense, it is important to stress that the research aims to focus on the *ethic cosmopolitans* (not ethic cosmopolitanism), reflecting on the biographical trajectories of the subjects (the volunteers) whose objective is the care for distant others, based upon an ideal of transnational solidarity. Under this vision, the social actor cannot 'escape' his/her moral duty regarding the constraints and destinies of the others but, at the same time, this orientation assumes that loyalty to a nation-state does not prevent the recognition of the self in a common humanity which is, in principle, the very foundation of the extension of solidarity beyond national borders (Cicchelli 2018). Following the classics of sociology, this approach towards solidarity can be easily confined in a precise relational dynamic characterized by coordination and mutual assistance among members of the same human *ensemble* – as social form of mutual recognition – making explicit reference to social ties, cohesion and social integration rather than to moral and legal shared norms. Solidarity therefore refers to a sort of reciprocity that creates a "conditional un-conditionality" that is linked to mutual recognition for the achievement of instrumental ends (Rosati 2001: 24). Following the contemporary sociological meaning of solidarity, it is easy to show that the semantic reference of the concept is broader, indicating a set of values and rules shared by members of a given society both with the purpose to act by coordinated actions – creating a mutual predictability (recognition of the expectations of behavior/role) – necessary for social integration and to refer to common tasks and a co-responsibility for the common good. In this sense, the building blocks of solidarity are: 1) values; 2) the shared rules; 3) a social space (or field of formal and informal relationships); 4) common goods; and 5) cooperative activities. (M. Rosati 2001: 19-20). The concept of solidarity – considered as a pillar of cosmopolitan socialization

– seems to be a very contradictory work in progress, also because the late-modern societal configurations – based upon contexts and social dynamics that are completely different from what we have seen in the past – imply a theoretical (and methodological) advancement in order to grasp new dynamics involving solidarity. In this sense, as highlighted by Oosterlynck (2014) late modern times require a renewed perspective in order to analyze the concept of solidarity[4]. In particular, the proposed conceptual framework considers *solidarity in diversity* as the most recent "stage of evolution" of solidarity after the rise and fall of the nationalization of solidarity. The authors suggest a significant modification of the epistemological choices connected with the scientific study of solidarity in late-modernity; in their own words, «new forms of solidarity can be identified if we re-adjust our focus» (2014: 12). For this reason, this issue recalls the dialectic relationship between individualization and identification in the social identity formation dynamics.

The starting point on this issue is the continuous tension between the individual and society. Once this situation is settled *de facto*, one may share the position of those who consider contemporary sociological debate to be 'imprisoned' in an insoluble polarity. That can be carried back to those points of view regarding the theme of identity: either a) a position that reifies the concept, or b) a radical processualism (Colombo 2007: 18). Besides its critical aspect, both theoretical positions summarily regard identity as an univocal and self-evident fact.

In our view, in accord with Antonina Kłoskowska's (2007) theoretic formulation, identity is a process more than a goal and has temporary multifaceted value; it refers more to social role than self-awareness; it is obtained outside self-awareness, showing its microstructural and historical character of a bond with a wider community and its culture. In order to situate Kloskowska's theoretic position regarding both poles of that already determined meaning – reification and radical processualism – it

---

[4] "The DieGem research project" promoted and coordinated by the University of Antwerp; Centre on Inequality, Poverty, Social Exclusion and the City (OASeS) implements a theoretical work package in order to develop and operationalize a rethinking of *solidarity in diversity* by using the most recent insights on solidarity, diversity, community, learning, citizenship, and place in different disciplines. The main hypothesis is that citizenship practices in daily life settings of forced closeness are important moments of social learning in which new and innovative ways of nurturing cross-boundary solidarity in diverse societies are developed. For further information see http://www.solidariteitdiversiteit.be/diegem_en.php (last access: 28/06/20).

could be set in a locus medio between the two categories: a sort of reifying processualism.

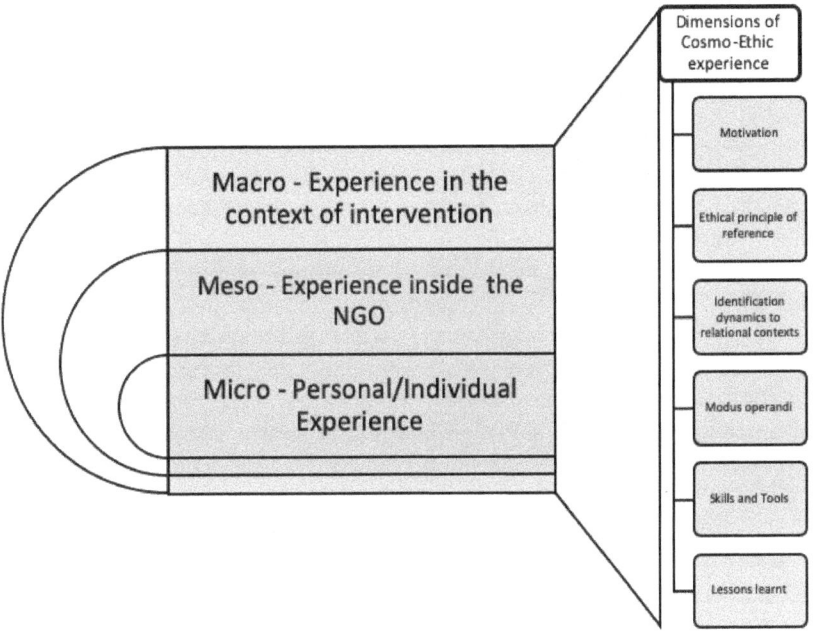

Fig.1 – Analytical model of the volunteering in developing countries as a cosmo-ethic experience

In fact, if that perspective considers identity as a process that is socially built and subject to temporal modifications, it is based, however, on a nucleus that is a thing (*res* in Latin), a culturally sedimented element; thus, it settles in a position that Brubacker and Cooper could define "hard essentialism" (2000: 1). In this sense, it is possible to argue that the concept of identity may be theoretically bent almost indefinitely. Identity – understood in a global sense – is an umbrella concept structured as a category of practices; therefore, it should be deconstructed in narrower analytic dimensions, which may become 'useful again' as they can be translated into concrete research tools. Following the theoretic-analytic route of the multipolar adaptation of identification (Gamuzza 2009: 136), it is more useful, for our analytical purposes, to replace the term 'identity' (a category of practices) with the term 'adaptation/identification dynamics to relational contexts'.

In order to blend the theoretical elements already briefly mentioned according to the heuristic focus of the research object – cosmopolitan solidarity practices and identity of the volunteers in developing countries working in NGOs – we present our theoretical model that, starting from Cicchelli's proposal about cosmo-ethic cosmopolitanism (2018: 246), finds its focus of attention in the discovery of how the cosmopolitan ethos interacts with three levels of the cooperation for development experience (personal-micro; within the international development cooperation body-meso; and within the context of intervention-macro). This theoretical articulation (motivation, *modus operandi*, virtues, emotions, and lessons learnt) was integrated – in consideration of the social unit under analysis (NGO volunteers for development) – with the "adaptation/identification dynamics to relational contexts" (avoiding the investigation of the emotional domain that in our view requires a specific research activity) in order to grasp the dynamic of learning processes connected with the relational dynamic in identification contexts. It was decided to add the category "tools" to explore the performative dimension related to the cooperation experience.

Bearing in mind these premises, during the theoretic phase I drew up a grid of analysis built by the sensitizing concepts highlighted by the sociological theories; this grid was used to build the research tools, proposing an exploration path to the researcher and, at a later stage, to interpret the information gathered during the research process. I always kept in mind the flexibility of research design in qualitative methods, in accord with Rubin and Rubin (2005: 44):

> adjusting the design as you go along is a normal, expected part of the qualitative research process. As you learn how the interviewees understand their world, you may want to modify what it is you are studying and rethink the pattern of questioning. Such flexibility is much better than persisting in a design that is not working well or that does not allow you to pursue unexpected insights.

Consequently, I left the door open to re-definition (refining) of the research tools. In this way, through the living experience (Ferrarotti, 1999: 7), we can underline the differences, tones and dissonances, that every biography unveils in a unique and personal fashion.

14.2 First evidences from the field and open-ended questions

The research project focuses on the in-depth analysis of ten life stories of international volunteers for development working for the NGO Co.P.E.[5] in Tanzania and Madagascar. The collected biographies cover a period of activity from 2004 to 2019.

As shown in table 1, the interviewees are prevalently female (7/10) with a high educational level (9/10 have a degree and one a PhD). Most of them currently live in Italy and the duration of their activity with CO.P.E. ranges from 1 to 16 years. The main roles during the experience abroad are project coordinator (4/10), international civil service (4/10) and country manager (2/10). At this stage of the work, it is possible to observe skilled subjects showing different levels of responsibility within the organisational structure of the NGO. All the interviews were carried out during face-to-face sessions (in the case of volunteers abroad it was decided to use Skype meetings) with a duration of at least one hour. The data collected have been analysed with QSR NVIVO© software.

For the sake of brevity, in this chapter a brief overview will be provided on the main emerging themes coming from the crossing of the analytical dimensions (motivation, ethical principle of reference, identification dynamics to relational contexts, *modus operandi*, skills and tools, and lessons learnt) with the three domains of reference of the cosmo-ethic experience (micro, meso, and macro) highlighted in the above analytical model; in this way it is possible to show how the experience of cooperation for development in Africa impacts different relational layers, generating a very interesting and alternative portrait of these experiences. The following presentation of insights will bring together the first three cosmo-ethic

---

5   The Non-Governmental Organization (NGO) CO.P.E. – Cooperation in Emerging Countries – is a non-profit and international voluntary organization, set up in Catania in 1983 and federated "FOCSIV – Volunteers in the World" (Federation of Christian Organisms of International Voluntary Service). Since 1988, the CO.P.E. has been recognized as suitable by the Italian Ministry of Foreign Affairs to carry out International Cooperation and Development Education activities (decr.n. 1988/128/5110/ID) in the context of Italian law n.49/1987 on Development Cooperation.
In Italy, the CO.P.E. carries out fundraising, information, and awareness and training activities in the field of Development Education (EaS) at national, regional, and local level in collaboration with other local associations. CO.P.E is also present in the following countries: Tunisia, Guinea Bissau, Madagascar, Tanzania and Peru, implementing projects in the health, agricultural, educational and social fields.

Table 1 – Profiles of Interviewees

Subject	Sex	Age	Marital status	Education	Religious affiliation	Current Work	Current Residence	Main role in ONG	Country of activity	Duration of activity (years)
1	M	46	Single	Degree	Atheist	Freelance	Madagascar	Project Coordinator	Tanzania	3
2	F	41	Single	Degree	Catholic	Italian Admission system (SPRAR) coordinator	Italy	Country manager	Tanzania Madagascar	16
3	F	38	Married	Degree	Atheist	Employee	Italy	International Civil Service	Madagascar	3
4	F	33	Single	Degree	Catholic	Program Officer UNICEF	Italy	International Civil Service	Tanzania	2
5	M	43	Single	Degree	Atheist	ONG Consultant	Italy	Project Coordinator	Tanzania	5
6	F	38	Single	Degree	Christian	Unemployed	Italy	Project Coordinator	Madagascar	1
7	F	34	Married	Degree	Christian	Educator	Switzerland	International Civil Service	Madagascar	1
8	F	37	Single	Degree	Catholic	Project coordinator Uganda	Uganda	International Civil Service	Tanzania	2
9	F	35	Single	Degree	Christian	Unemployed	Italy	Country manager	Tanzania Sud Sudan	8
10	M	35	Single	PhD	Catholic	Research	Italy	International Civil Service	Tanzania	1

dimensions – the ideal components of the model – and the three operative dimensions (*modus operandi*, skills and tools, and lessons learnt) to report the performative and practical side of the collected biographies. In this way, even if in a form of a *résumé*, it is possible to draw some provisional conclusions regarding the stage of advancement of the research.

Table 2 – Synoptic scheme of the main themes emerging from the fieldwork

Experience Domains / Cosmo-Ethic Dimensions	Micro – Personal Experience	Meso – Experience in international development cooperation body	Macro – Experience in the context of intervention
Motivation	Care for the others; Support by significant others; Career opportunity and individual development	Reduction of inequalities leaving time for the subject to adapt to the local context	Perception of the trajectories for the development of the context of intervention
Ethical Principle of reference	Solidarity towards distant others	Social responsibility of the organization	Mutual Recognition with local communities
Identification dynamics to relational contexts	Family/friends interactive field	ONG organization mission as way of life	External Local community
Modus operandi	Involvement/ participation	Project-based intervention	Active involvement of local communities
Skills and Tools	Individual background skills	Relationships with the organizational-administrative structure of the organization	Realization/ management of tangible (structures and services) and intangible (knowledge; know-how etc) assets
Lessons learnt	Relational/technic-operative skills	Competencies in organizational innovation (based on individual experience in a bottom-up approach)	Skills in Development/ underdevelopment of the context after ONG actions

The motivation issue, at micro level, showed itself to be a very important domain for the definition of the experience. Our respondents revealed a very interesting variety of motivations (crossing the three domains, micro, meso, and macro) ranging from: a) very contextual decisions based upon an instant evaluation done during the application phase as declared by a Madagascar civil service volunteer:

> Then the project was new, mmh ... when I applied there was already a volunteer who had been there [Madagascar – ed.] for six months, it was all in the bud, as after all I was also a new volunteer at that moment and I said to myself "maybe this could be right for me" also because in Tanzania anyway I had seen that there were projects that had nothing to do with my studies ... I liked the idea of Madagascar, then I also considered I might have more possibilities [of being chosen.] for the project. There are many English speakers [during the selective interviews – ed.], maybe French could be a point in my favour since it compensates for the lack of experience. I have to say it was a good choice because it was a wonderful experience for me. I grew together with the project (int. 3, F, Madagascar),

to b) more meditated choices related to the educational paths: half of the interviewees (5/10) have a degree in international relations or political sciences with an international curriculum. The others, (architecture, forest sciences, agricultural sciences, and education sciences) put in practice their competences as a leverage for motivation through the project coordination from a technical point of view. In this sense, the interviewees showed a very high motivation for shaping their biographical experience with a period abroad as a sort of biographical *rite de passage* (Turner 1972): a natural evolution/consequence of the previous choices.

> Yes... from Syracuse (city in Italy, Sicily – ed.) exactly. Then I spent a year in Ireland. I did this internship at Oxfam [NGO – ed.] and after that I enrolled in Lyon university in France and I did a degree in human rights, so I did this two-year II level degree course...yes... two years. The second year I moved to Spain because I did the internship there in the field of immigration with an NGO that deals with immigration in Andalusia, especially between Morocco and Spain, and among other things before, the summer, so 2011 I started in Tanzania with CO.P.E. (int. 7, F, Madagascar)

In addition to this, another intriguing issue emerged from the first wave of analysis: the relevance of the support from significant others. In particular, the family background/approach to solidarity represents a positive (or negative) catalyst in the decision to volunteer abroad to take care of distant others. Solidarity is, without any doubt, the ethical principle of reference placed by our interviewees in connection with its concrete realisation in the ONG goals and methods for assuring social responsibility.

The aim of the dimension of identification dynamics to relational contexts is to explore, through biographical experience, how the respondents were able to deal with the self-identification conflicts (and troubles) in creating (or re-shaping) a coherent identification process through the elaboration of the practices/experiences. The micro domain, pertaining to primary groups, reports the relevance both of the family – even if there are some critical issues to overcome – and friendship circles as relevant identifying contexts. As highlighted by one of the project coordinators regarding family obstacles:

> […] the family let's say, at a wider level, they have always accused me of being very selfish, that is, although my choice for some could be, how can I say, altruistic, oh well it's a mix of the two, however they, at the family level, have always accused me of being selfish, not thinking about the concern that I brought to my family, to everyone in short, because I live abroad, but all in all, over time, they had to adapt to my choices (int. 6, F, Tanzania)

In other terms, our respondents reveal an "identification bargaining" between, on the one hand, the personal transformative experience co-created in interaction with the local communities in Africa and, on the other hand, with the NGO interactive circles of reference in the home country. The relevance of primary interactive circles – and its elaboration as identification context – can be seen through a fragment of experience of a civil service volunteer working in Tanzania for one year. When he came back to Italy after one year, the initial difficulty for him was telling the story of his experience. The turning point is represented by the shared elaboration of the experience during a meeting with other colleagues he worked with in Africa during a meeting in Rome. Moreover, it was necessary for him to have a period of time to find a way to re-elaborate the relationship with friendship circles under the light of the volunteering experience:

> … my friends, when I came back, there were already those who said "when are you coming to tell us about it? when are you coming to talk" [about your experience - ed.] so in the end I was very tired, suffering from a very strong emotional charge. At the same time, I was happy to come back, because after eleven months you are happy to come back, but I had no desire to meet anyone or tell anything because it was as if that experience was mine, it was an intimate and private thing. But I must say that the post Tanzania period was one of the best in my life, because there were two or three months of light-heartedness and finding myself with all that I had left before. I had a post-experience meeting with the colleagues who had been with me. We met and there I began to assimilate, to recount, to participate in a series of events, seminars and activities in which I gave voice to my experience. (int. 10, M, Tanzania)

The dimensions of the *modus operandi*, skills and tools, and lessons learnt explore the operative-oriented components of volunteering as a cosmo-ethic experience. The main narrative outputs undoubtedly show the extent to which volunteering requires a significant 'shift' in the personal approach to life, time, and the nature of relationships:

> The first thing is the concept of time, because it is completely different, in many aspects: one, the daily time, that is, the work starts at 8 but people do not arrive at 8 because time is not measured ... it is measured with "the sun is high, the sun is low". That is, there is a completely different concept of time and through a series of suggestions and out of curiosity, I found out that this thing is written in Ebony, by Kapuściński, in one of his books [...]. Now, the concept of time, here in Europe, in the western world, is completely different from that of the African world. [...] The time here, for us, in the western world is an entity in its own right ... time passes regardless of what your work is, but in the African continent time occurs only when you make it manifest, so if you do nothing, you do not have that feeling of anguish of wasting time because at that moment time does not manifest itself and does not exist, and is perhaps the key that allows it to go to what I initially considered a sort of "brain stand-by" because you see them and you see that they are in absolutely no hurry, no kind of stress in seeing that half a day has passed and nothing has happened; for us, this thing ..., especially when you have goals, always because you act as a link between these two worlds that still have time in between, a project that begins and must end and a temporal projection must be there when it ends, not to mention the stuff of projects with financiers who give you a priori the midnight of a day after two years of project ... this is one of the elements of greatest stress for the volunteer, like me. (Int. 1, M, Madagascar)

These words show effectively that the modalities of involvement and participation (micro domain) have been re-interpreted in the light of a new frameworks of reference given by the experience. The project-based intervention (meso domain) becomes a new interactive realm in which everything is filtered by new cultural codes of reference in which the subject acts. In this way, the only way to deal with the new reality (macro domain) is to find an efficient way to cope with the local community through active involvement. This point is relevant even considering the lessons learnt on both sides of the relationship. It is obvious that this element cannot be underestimated in every relationship (especially in work-mediated relationships), but our interviewee testifies the need for a soft way of dealing with people to succeed in this task:

> I often found myself, for example in building, when you prepare a construction site you have to start plotting and there are different ways of doing this. If you know the geometry, you can manage angles, bisectors and tangents, a whole series of things, that you could do it in a simpler way or in a slightly

longer traditional path. I used to play with this a bit because it was fun, because it also gave me a certain type of recognition within a society that has this crackhead saying to you: "the point in our opinion is here". We often took these measurements – and obviously knowing the geometry you got there in the shortest way – and then as soon as I turned around I realized that they had retraced the points in their own way to get to that point and after some time wasted in checking my results, which was not wasted, because it served to realize this form of mutual recognition and say "ah, then probably you also know something too!". Obviously when you have good relationships, you tease each other as well. I often had to test my physical endurance in ways that I would never have done here, simply because there I couldn't fail. I had to carry the planks on my back as they did, I had to transform my body, because if they had to do it, they did it and then they came running and said: 'You know something? You were right'. This did not mean that the subsequent plotting was done in one way, that is, it was done anyway, because it was also right to verify, to verify. (Int. 1, M, Madagascar)

Skills and tools and lessons learnt can be considered as two faces of the same coin. These two dimensions were required, in the words of the respondents, to be re-defined in the light of the relationship with the cosmo-ethic solidarity experience translated by the relationships with the three domains (individual, organisational, and contextual) impacting the biographical route. The individual skills, especially at micro level, represent the background of experience but this is not enough; they are not only technical data but, most of all, relational. To be clear, the necessity to deal with the organisational-administrative structure of the NGO (meso domain) does not require only technical (albeit important) competences but, most of all, the ability to manage a distant relationship with a very strategic context for personal equilibrium. Nevertheless, this process is not without contradictions, as declared by one of the country managers with a very long experience in the NGO (16 years) that tried to find an exit strategy (as a lesson learnt from the direct experience) to this situation, proposing an innovative organizing model for CO.P.E. in order to promote the active participation of the volunteers abroad in the life and strategic choices of the NGO:

I pointed out that during the assemblies [general annual assembly of the NGO – ed.] it was as if there were no volunteers down [in Africa – ed.] sometimes, and this thing really annoys me, that is: I would like to see the faces of the volunteers, for them to be "present" during the assembly because it is right for many reasons, because the volunteers down [in Africa – ed.], however, help to make up for shortfalls or other things or in any case provide added value, for better or for worse, which must be taken into consideration, must be discussed and reported. I remember that it was like that back then, we reported all these things...but what about the situation in Italy? That is, for me: okay I came back once but for two months, and I was able to pick up the communication lines here in Italy...what they are doing and that gives me strength, but sometimes I feel between a rock and a hard place as the country manager, having to liaise between here and there, because you have volunteers on the one hand who tell you a whole series of situations, even amplified, even exaggerated sometimes, and on the other you have the Italian office [of the NGO] that sometimes does not understand what is going on down there. (int. 2, F, Tanzania)

In order to draw some preliminary conclusions of the work already done it seems necessary to divide our final remarks into two parts: theoretical and operative. The interpretive model proposed in this work developed the cosmopolitan socialization theory proposal for designing a research path with specific social units. The original model was integrated with the dimension of identification contexts in a relational context. Up to now, the heuristic model revealed itself to be informative in combination with the use of biographical data. Thanks to the richness and density of it, the idiographic approach enabled the researcher to preliminarily enter the 'black-box' of the subjects revealing a very composite process related to the cosmopolitan socialisation dynamics activated by the volunteering experience. It is interesting to note that the osmotic relationship between the different domains (micro-macro of the experience) affected the whole process in a karstic way. In any case, a theoretical saturation of the model seems necessary thanks to additional cases to be considered. In addition to this, the biographical research needs to be combined with the quantitative perspective, since the horizon of uniqueness and the depth of the individual experience is the horizon from which to discover the sociological sense of the phenomenon and its effects. This research experience can be useful not only on the scientific-academic side: the research results can be useful to design a rigorous system of validation of cosmopolitan competences for the international volunteers in developing countries; in this way it will be possible to suggest how they can best contribute to their country of origin during the periods they stay in Italy waiting for their next "cosmopolitan-ethic experience", or if they decide to remain permanently in the home country.

Bibliographical references

Bhambra, G. K., 'Sociology after Postcolonialism: Provincialized Cosmopolitanism and Connected Sociologies', in M. Boatcă, S. Costa and E. Gutiérrez-Rodríguez (eds), *Decolonizing European Sociology: Trans-disciplinary Approaches* (Aldershot: Ashgate, 2010), pp 33-47.

Beck, U., 'The Cosmopolitan Perspective: Sociology of the Second Age of Modernity', *British Journal of Sociology*, 51 (1) (2000), 79-105.

Beck, U., *Cosmopolitan Vision* (Cambridge: Polity Press, 2006).

Beck, U., and N. Sznaider, 'Unpacking Cosmopolitanism for the Social Sciences: A Research Agenda', *The British Journal of Sociology*, 57 (1) (2006), 1-23.

Beck, U., 'Risk, class, crisis, hazards and cosmopolitan solidarity/risk community - conceptual and methodological clarifications', FMSH-WP-2013-31. 2013. <halshs-00820297> (2013).

Brubacker, R., and F. Cooper, 'Beyond "Identity"', *Theory and Society*, 29 (2000).

Cicchelli, V., *Plurale e comune. Sociologia di un mondo cosmopolita*, (Roma: Morlacchi, 2018).

Cipolla, C., *Oltre il soggetto per il soggetto. Due saggi sul metodo fenomenologico e sull'approccio biografico* (Milano: FrancoAngeli, 1993).

Chouliaraki, L., 'Cosmopolitanism', in J. Gray and L. Ouelette (eds), *Media Studies* (New York: New York University Press, 2016).

Colombo, E., 'Decostruire l'identità. Individuazione e identificazione in un mondo globale', *Culture*, 19 (2007).

Ferrarotti, F., *L'ultima Lezione* (Bari-Roma: Laterza 1999).

Gamuzza, A., *Identità al confine. Concetti teorici e ricerca empirica* (Milano: FrancoAngeli, 2009).

Held, D., and P. Maffettone, 'Moral Cosmopolitanism and Democratic Values', *Global Policy*, 8 (6) (2017), 54-64.

Kłoskowska, A., *Alle radici delle culture nazionali* (Reggio Emilia: Diabasis, 2007).

Leenstra, M., 'The human factor in development cooperation: An effective way to deal with unintended effects', *Evaluation and Program Planning*, 68 (2018), 218-224.

Melucci, A., *Parole chiave. Per un nuovo lessico delle scienze sociali* (Roma: Carocci, 2000).

Macioti, M. I. (ed.), *Biografia, storia e società. L'uso delle storie di vita nelle scienze sociali* (Napoli: Liguori 1985).

Oosterlinck, J., Loopmans, M., Schuermans, N., Vandenabeele, J., and S. Zemni, 'Putting flesh to the bone: looking for solidarity in diversity here and now', DieGem working paper, retrived in http://www.solidariteitdiversiteit.be/uploads/docs/bib/integratiepaper.pdf [accessed 28 June 2020].

Putnam, R., 'E Pluribus Unum: Diversity and Community in the Twenty-first Century The 2006 Johan Skytte Prize Lecture', *Scandinavian Political Studies*, 30 (2) (2007), 137-164.

Rosati, M., 'La solidarietà nelle società complesse', in F. Crespi and S. Moscovici (eds), *Solidarietà in questione* (Roma: Meltemi, 2001).

Rubin, H. J., and I. S. Rubin, *Qualitative Interviewing (2nd ed.): The Art of Hearing Data*, 2nd edn (Thousand Oaks: SAGE Publications, Inc., 2005).

Skrbis, Z., Kendall, G., and I. Woodward, 'Locating Cosmopolitanism:Between Humanist Ideal and Grounded Social Category', *Theory Culture and Society*, 21 (6) (2006), 115-136.

Skrbis, Z., and I. Woodward, 'The ambivalence of Ordinary Cosmopolitanism: Investigating the limit of Cosmopolitan Openess', *The sociological Review*, 55 (4) (2007), 730-747.

Turner, B., 'The enclave society: toward a sociology of immobility', *European Journal of Social Thoery*, 10 (2) (2007), 287-304.

Turner, B., *Il processo rituale* (Brescia: Morcelliana 1972).

Conflict in Education

GIOVANNA DA MOLIN, ANGELA CARBONE[*]

15.
THE ITALIAN FAMILY:
HOUSEHOLDS, RELATIONSHIPS AND CONFLICTS
(XVIIth-XIXth CENTURIES)

15.1 *A historical perspective on the family*

15.1.1 *Family structure and size*

Studies on the family have been undertaken in many different subject areas including history, sociology, anthropology, philosophy, economics, law, theology and demography.

Different methodologies, source materials, criteria and perspectives on the transformation of the family over the centuries have led to lively scientific debate, especially between historians and sociologists (Saraceno and Naldini 2013; Caporrella 2008).

Pioneering studies by renowned sociologists in the 19th century, in the transition period from traditional society to the contemporary society of the time, maintained that the format of the domestic household was becoming progressively simplified, moving from the multiple family, consisting in different generations and married couples, towards the nuclear family of couples and their children. Émile Durkheim stated the "law" of the progressive contraction of the family and, prior to that, Frédéric Le Play had argued that the 'patriarchal' family and the family of origin were being replaced by the "unstable family", i.e. the nuclear family. According to Le Play, this was due to industrialisation and urbanisation as well as to the changes in inheritance laws (Le Play 1864: vol. I 193-329, vol. II 44-60; Bernardi 1981).[1]

The contributions of historians in the mid-1960s, who had begun to study the family as a specialist area of study, radically contested the idea that the

---

[*] The layout of the chapter was organized by both authors. Specifically, Giovanna Da Molin drafted sections 15.1 and 15.2, Angela Carbone drafted sections 15.3, 15.4 and 15.5.
[1] For a detailed point of view about the "law of contraction" of the family, please consult: Émile Durkheim, *Les règles de la méthode sociologique* (Paris: Alcan, 1910); *De la division du travail social* (Paris: Alcan, 1911); *La sociologie* (Paris: Larousse, 1915).

era of industrialisation was the great watershed between the traditional and nuclear family, thus paving the way for numerous new studies on the history of the family and for new interpretations (Stone 1987; Anderson 1982).

The publication in 1972 of *Household and Family in Past Time* by Peter Laslett and Richard Wall was decisive from a demographic and quantitative point of view. It presented the research results of the *Cambridge Group for the History of Population and Social Structure*, founded by Laslett and his student Tony Wrigley in 1964 (Wrigley and Schofield 1981).[2] The main aim of Laslett and his co-workers was to study the British family in the preindustrial era, concentrating on the size and composition of households.

They began by reconstructing the organisation of the family based on the available sources. In order to do so, such sources must contain specific information, such as the registration of inhabitants in a given place, in a given year, grouped by nuclear family, as well as the degree of family relationship and cohabitation. Other elements that can contribute to historical research are the types of employment and socio-economic conditions of its members. Sources that correspond to these criteria are, for example, lists of names of the population, such as the *listings* in England or the *status animarum* of Catholic Europe, or even economic sources such as the 17th century *catasti antichi* (real estate registers) and the 18th century *catasti onciari* of Southern Italy (Da Molin and Carbone 2016).

The most important aspect of the research by Laslett and the Cambridge Group was the elaboration of a classification system which could be used by historians and demographers in other countries, facilitating both territorial and chronological comparison (Laslett 1972: 847-872, Italian translation Barbagli 1977: 30-54).

Laslett's classification is based on the simple family unit of married couples with or without children, or widowers or widows with children. The expressions *domestic household*, *nuclear family*, *simple family* or *biological family* all indicate the *married family unit*. Apart from the *nuclear* or *simple family*, the classification identifies five other types of family structure.

The *extended family*, or extended household, is a nuclear family plus one or more relatives who do not form other couples. If the additional members belong to an older generation than the head of the family, for example, a widowed parent, an uncle or grandfather, it is classified as an upwards

---

2   Initially, the study of the domestic group was only one aspect of a larger research program on the English social structure that had to be analyzed in close correlation to a study on the demographic characteristics of the English population in the long term.

extended family. If however, they belong to a younger generation, then it is a downwards extended family. If the additional member belongs to the same generation as the head of the family, for example, a brother, sister, sister-in-law or cousin, it is a laterally extended family. Some nuclear families can be extended upwards and laterally or upwards and downwards, and so on.

The *multiple family*, or multiple household, is one with more than one married couple who are closely related, living under the same roof. They are classified depending on the generational direction: upwards, downwards, on one level, *frérèches* and so on. Second unit upwards families are those with an additional member or secondary unit from an older generation (e.g. parents who live with a married son or daughter). If, on the other hand, a couple and their children live with their parents then the family is a second unit downwards family. If the family unit is horizontal (e.g. a widowed parent plus two or more of their married children), then it is on one level. The French term *frérèches* is used to describe a family where there is more than one married sibling but no coresident parent. Other multiple families include households with more than one nuclear family, related or not, who belong to different generations from the head of the family.

Non structured or *No family* refers to households with no married couples, i.e. people who may or may not be related but live together, e.g. unmarried siblings, uncles, aunts, nephews or nieces or unrelated co-residents with an implicit or explicit mutual interest.

Solitaries are people living alone, widows or widowers with or without domestic help.

The classification also includes a sixth category of *Indeterminates*, i.e. those whose relationship with the head of the family is not indicated in documents or to whom it is impossible to assign a classification due to the poor conservation of the sources. In the 1960s, numerous studies on the structure of the family in pre-industrial Britain were made by Laslett and his co-workers using this classification, with surprising results.

The results showed a long-term stability in family structure and size. In the pre-industrial period between the 16th and 18th centuries, the most common type of family unit in England was the nuclear or simple family, which was relatively small with an average of 4.75 members.

Contrary to Le Play's theory, which identified industrialisation as the cause of smaller family size, Laslett's research results showed that the increase in multiple families in England occurred after the beginning of the Industrial Revolution. According to Michael Anderson, this phenomenon was linked to the economic advantages gained from living with parents or other relatives, especially in the early years of industrialisation and urbanisation, thus favouring the creation of complex families (Anderson 1971).

Laslett was heavily criticised for his classification, with Lutz Berkner openly disputing his methodology and the static approach adopted. According to Berkner, family census occasionally recorded family groups at various stages in their development. The family should not be viewed as a static reality but as a constantly changing one. Births, deaths and marriages consistently produce changes in the family structure and its number of members. Moreover, growth and old age change both the roles and the attribution of authority and power among the different family members. Therefore, a household can be made up of a nuclear family at one particular period in time, but also of an extended or multiple family at different stages in its development.

Berkner suggested analysing the distribution of households according to the age of the head of the family. He maintained that for each of the different age groups of the head of the family, each family type would show different frequencies and that the dominant structure in each group could be considered typical at a specific stage in its development. Moreover, the low percentage of multiple families registered in pre-industrial England could be correlated to the high rates of mortality often linked to marriages between older people, factors that meant a family could have a complex structure only for relatively short periods within its evolution (Berkner 1972: 398-418; Id. 1975: 721-738).

Without doubt, in each type of family, especially in the past, the age at which people married was an important variable, as widely documented by the English demographer John Hajnal, whose name is linked to an important article from 1965 (Hajnal 1965, Italian translation Barbagli 1977: 267-312).[3] Hajnal demonstrated that for centuries marriage in Western Europe was characterised by the older age of the partners of both sexes (26-27 for men, 23-24 for women) as well as the high number of unmarried people (10-15% of the population).[4] Conversely, in Eastern Europe, early marriages were typical and almost the norm both for men and women. Later studies showed that it is not possible to define one singular model of marriage in such a clear-cut fashion. However, both areas of Europe had their exceptions: for example, specific studies have shown that in Ireland, Southern Spain, Sicily

---

3   Furthermore, Hajnal is the creator of a method for calculating the average age at first marriage based on the distribution of a population by age, gender and marital status. See: Hajnal 1953.
4   Regarding some aspects of unmarried people in modern Europe and the international bibliography, see: Lanzinger and Sarti 2006.

and Apulia, people married at an early age, and this was greatly influenced by social class, especially for men. The differences in age at marriage also determined a different framework of the family structure. In Northwest Europe, for instance, young married people went to live on their own, thus forming a new family unit with the husband as head of the family. In Eastern Europe, however, young married couples continued to live with their parents, usually with the husband's family, thus forming a multiple family. Moreover, the average age at marriage and the various types of family structure that characterised pre-industrial Europe were influenced by the custom of having domestic servants in the household. According to John Hajnal, in the simple families of Western Europe, people married later because young couples were expected to create an independent household and therefore needed to have sufficient income to "set up home". Consequently, young people worked as servants before getting married.[5]

Apart from some exceptions, young people in Eastern Europe did not work as servants for other families before getting married. With regard to Italy, from the early 20th century scholars had already begun to study marriage and the family. In 1915, the demographer Livio Livi published a study on the variations in time and space of its size and composition (Livi 1915). However, the most significant input on the history of the family in the past was provided by legal historians, including the contribution of Nino Tamassia in 1911 on domestic relationships in Italy in the 15th and 16th centuries (Tamassia 1911).

Italy has very rich sources on the history of the family structure. These list the inhabitants of a community for a specific year, grouped by family, with information on the types of family relationships. These are to be found in, for example, religious sources such as the *stati delle anime* and civil and tax sources such as *le consegne dei capi di casa* in Piedmont, the *riveli di beni e di anime* in Sicily and the *catasti antichi* (real estate registers) and *catasti onciari* in Southern Italy, and are conserved in ecclesiastical and public archives throughout the country. Yet despite this substantial

---

5 Regarding the history of household servants, please consult: Hajnal 1983: 65-104, Italian translation Wall, Robin and Laslett 1984: 99-142; Reher 1998: 203-234; Sarti 2006a: 3-59; Sarti 2006b: 222-245; Clegg 2015: 43-66. For a critical discussion of the Cambridge Group studies, see: Guttormsson and Viazzo 1988: 355-379; Hinde, Ago and Torrente 1988: 541-571. Authors who offer a lucid overview on household servants in early modern Italy are: Barbagli 2013; Arru 1992: 273-306; Ead. 1995; Da Molin 2000: specifically 191-215; Ead. 1992: 219-252.

wealth of documents, studies on the family began later in Italy than in other European countries.

The first epic work on family structure in Italy was carried out by David Herlihy and Christiane Klapisch-Zuber in 1978, based on the extraordinary source of the Florentine land registry of 1427 (Herlihy and Klapisch-Zuber 1978, Italian translation 1988). The authors reconstructed approximately 60,000 families, with a total of 260,000 individuals, with an average of 4.3 members per family unit. The distribution by family type, according to Laslett's classification, showed that 55% were nuclear families, 10% of the total number registered were extended families, and 19% were multiple families, with a considerable number of solitaries (14%), and a reduced number of no family (2%). This study highlighted a significant difference between towns and countryside in 15th century Tuscany, showing a high proportion of multiple families in the rural share-crop population.

Innovative studies include those of sociologist Marzio Barbagli, who studied the changes in families in northern-central Italy between the 15th and the 20th centuries, the anthropologist Gérard Delille who studied some areas in the Kingdom of Naples between the 15th and the 19th centuries, and Giovanna Da Molin who studied the demographic-quantitative aspect of families in modern Italy.[6]

15.1.2 *Relationships and sentiment*

So far, this paper has dealt with studies on the family in the past, focusing on its structure, i.e. the coresident household. However, the family, as such, inevitably begs other important questions that have given a new slant to historical research based on new sources and documents. This does not lessen the interest in coresident family structure but rather introduces different perspectives which consider the fabric of relationships within the family (relatives, family strategies, alliances, friendship etc.) and the authority and affection that exist within the coresident family group.[7]

These topics have been dealt with to a lesser extent when compared with the research on the quantitative-structural aspect of households that took inspiration from Peter Laslett and the Cambridge Group. The reason is that sources are often difficult to find and to study.

---

6   Barbagli 2013; Delille 1985, Italian translation 1988; Id. 2011; Kertzer and Saller 1991; Casanova 1997; Da Molin 2000: 47-134; Raffaele 2000.
7   Hareven, Kern and Plakans 1987; Wall, Hareven and Ehmer 2001; Sovič, Thane and Viazzo 2016; Sarti 2017.

With regard to family relationships, in particular, those with close relatives, the most important documents are notary files, as indicated by Giovanni Levi in the 1980s (Levi 1985a; Id. 1985b). For example, in marriage agreements which were drawn up when a dowry was compiled, or in wills and testaments, an intricate network of family relationships, alliances, conflicts, strategies and behaviours was revealed which were quite diverse depending on geographical location and on the different social classes that formed the populations of the past.

Within the network of family relationships, spiritual relatives (e.g. godfathers and godmothers, bridesmaids and best men) provided further support (Alfani 2006; Alfani and Gourdon 2012a: 1005-1028; Alfani and Gourdon 2012b).

By analysing the authoritative and sentimental rapport between husband and wife, or parents and their children in the family of the past, historians addressed the problem of the breakdown of the structure and organisation of the traditional family and its transformation into the intimate modern one that was formed between the 18th and 19th centuries. A characteristic of the traditional family was its extreme vulnerability to external influence and community control. The relationships between different members of the nuclear family were deferential, authoritarian and imposed by the head of the household. Marriage was seen as a "contract", mainly of an economic nature, for productive and reproductive purposes. It was a mechanism for passing on the family wealth from one generation to the next, with little or no affection. Slowly and gradually this family model was replaced by a more intimate one which was more liberal in the distribution of authority and built on the affection between spouses and parents and their children rather than on financial interest.

Lawrence Stone maintained that, in England, the simple family emerged in the second half of the 17th century in the upper classes of English society. Greater consideration was given to the marital relationship and raising of children at the same time as the development of a market economy and, consequently, social mobility (Stone 1979).

Edward Shorter, however, associated a wave of true sentiment with the beginning of the Industrial Revolution, especially between workers. This was a sort of sexual and romantic revolution, resulting from a less repressive culture and the separation from the original family because of urbanization (Shorter 1976).

If the substitution of deferential language forms by the more informal form of "you" (*tu*) is to be considered significant, then studies by Marzio Barbagli on the correspondence of noble families in Italy show that the move away from the traditional patriarchal model of the family to the new intimate one began with noblemen born in the last thirty years of the 18th century (Barbagli 2013).

Despite the significant contribution that studies on the history of the family have made to the understanding of social mechanisms in the modern age, many questions remain unresolved, and historical research in the last few years has been making use of the numerous stimuli, detailed research and different study perspectives.

Going beyond the interpretation of the family structure, microanalytical studies consider how the family was viewed from within. The analysis of this microcosm focusses on the role of each individual member and their mutual relationships. In this context, the value of intergenerational and intragenerational relationships plays a decisive role. It is, in fact, the intergenerational relationship that links the institutional and interpersonal dimension. The family, as part of the generational fabric from which it was created, is the origin of every social relationship and a member of the society to which it belongs. The family is the mediator of the sexes, the generations and the individual and society: it is the place where each individual is defined in relation to gender, their place on the generational ladder (parent and/or child) and their stage in the life-cycle (age). Therefore, the generational perspective facilitates the interpretation of family ties as a combination of generational histories that continually intertwine.

Over the centuries, the relationship between generations greatly influenced not only the sentiments but also the structure of the financial resources of a family, their helpers, and the symbolic and political representation of their ideas (Fazio and Lombardi 2006; Ago and Borello 2008; Casanova 2009).

15.2 The Italian family in the 17th and 18th centuries: structure, size and extension

Research on family structure in the past in different regional areas in Italy has revealed a mixture of situations and circumstances.

The structure of the family was influenced by numerous factors, such as: demographics (marriage, average age when first married, death, migration); land and population distribution (differences between town and country); economic and social factors (job type, wealth, the handing down of trades from father to son, servants); and religious and cultural aspects or more deep-rooted ones such as sentiment, family strategies and inheritance.

Despite specific differences throughout various areas, the Italian family in the 17th and 18th centuries was a simple family comprising parents and a modest number of children (Da Molin 2014: 65-111).

Simple families were more prevalent in cities. Compared to rural communities, the cities also registered higher percentages of solitaries; for example, young people in search of work or better fortune, elderly widowers or widows who had no family support and those who had joined religious orders, as well as households with no family structure, which included unmarried relatives or unrelated coresident persons living together by choice or necessity. Without doubt, the most significant figures were registered in the papal city of Rome, where this type of family accounted for 15% of the total registered in the mid 17th century. Therefore, there were fewer extended or multiple families.

In rural communities, family models were closely linked to the various types of community and the organization of the land that characterised Italian rural areas in the past.

From a schematic point of view, it can be observed that rural families were generally simple, with a small percentage of complex households in areas with high numbers of farmworkers. A perfect example is the region of Apulia, which we have studied in detail, that showed a high prevalence of simple families as, after marriage, a high percentage chose to set up separate households.

Throughout the modern age, the percentage of simple families was never less than 65% in the Apulian communities, at times reaching 80%. At the same time, the percentage of multiple families remained at a modest 2-3% of the total.

Regardless of geographical position, the tendency to set up simple families independent of the original family was characteristic of agricultural labourers with no privileged ties to the land. The high percentage of nuclear families registered was not exclusive to farmworkers in Southern Italy but extended to the *brazanti* and *pigionanti* in the flatlands south of Milan, the farmworkers around Vicenza, and the *pigionali* in the countryside around Prato.

Conversely, the widespread systems of tenant farming and crofting found throughout the vast areas of northern-central Italy tended to produce more complex and larger families. Where the rural family constituted a family business, couples who continued to live in the original family home after marriage, co-residents of more than one family group and scattered settlements were all part of the systems and organisation of work and production. In the large families of tenant farmers in the areas of Piedmont,

Veneto, Tuscany and Emilia-Romagna, the logic of living with relatives under the same roof was replaced by that of the necessity to produce while, in order to deal with the danger of proletarianism, the tenant's main aim was to maintain the balance between producer (that is, the workforce) and consumer (mouths that had to be fed).

While in the rural communities the demarcation line was between stable and unstable resident groups, sharecroppers, owners and agricultural workers, in the city structural changes in the household were determined by the distribution of wealth.

The tendency to live in multiple and bigger families corresponded to higher social and economic status. Families of noblemen and landowners were more frequently extended to include unmarried relatives or sometimes married couples from the same family. The ancestral family home, a tangible sign of power and cohesion, was the centre of family life for the large aristocratic families, while having large numbers of servants was a sign of prestige.

15.2.1 *The extended family*

Within this extremely complex mix, it is difficult to define one single model for the traditional family in the past. It was not unusual for a simple family to include other relatives. Studies on family ties between co-residents show different behaviours and relationships. The reasons for extending a family were either financial or linked to sentiment or solidarity. In some areas in northern Italy, relatives were accepted into families to make up for the lack of workers. In Villafranca, near Turin, in 1622 the family was extended to include members from the same generation as the head of the family, following an exclusively male line.

Another reason was purely for solidarity. The family was extended upwards to include an elderly widowed parent.

In many areas of Southern Italy, the simple family structure changed if the head of the family's widowed mother or mother-in-law moved in, often accompanied by unmarried or widowed sisters or sisters-in-law. The son responsible for continuing the family line was obliged by specific cultural codes to welcome and care for his mother, sisters, sisters-in-law and unmarried female relatives by providing moral and financial support.

Recent research within a diverse and variable framework is focussed on highlighting the solidarity between generations, the importance of the horizontal family, the ties between young couples and their original families and the relationships between parents and their children. Although the family structure tended to be simple, from the end of the 18th century in Italy and in other European areas, these relationships would pave the way for a more intimate family, more liberal than patriarchal in the distribution of authority and closely bound by sentiment between spouses and between parents and their children.

Moreover, in the past, while intergenerational relationships played a decisive role within the family, they were also important for the support system surrounding it with regard to the weaker relatives: children, the elderly, single women, the sick or isolated.

15.2.2 *Household size*

In general, regardless of geographic location, the Italian family in the 17th and 18th centuries was relatively small, with size and structure correlated. In both the North and South of the country there were, on average, 4 to 5 members per family, and rarely more than 5.

There was a clear distinction between urban and rural families, urban families being much smaller in most areas of northern-central Italy. In the city of Moncalieri (Turin) in Piedmont the average number of family members was 3.8, while in the rural areas this increased to 4.9. The same difference was observed in the city of Carmagnola, in the same area, where families residing within the city walls averaged 4.7 members, while those in the surrounding rural area had an average of 5.4.

Equally relevant were the trades of the heads of the household, their wealth and social position. For example, in Perugia in the mid-17th century, the parishes in the town centre where most of the wealthy and professional people lived registered an average of 5 members per family. Conversely, in the suburban parishes where more humble families lived, the average number was 3 members. Similarly, the aristocratic families of Pisa, who were wealthy landowners, merchants and businessmen, registered an average of 5 members per family, while the workers and apprentices had an average of 4 members.

In Southern Italy, wealthy families also tended to be bigger compared to the lower classes. A better quality of life meant a higher possibility of survival for the children of noblemen and landowners, while the lower classes registered a high level of infant mortality. Moreover, noble families often had servants who did not change the structure of the family but did affect their size. Noble families in Bari in the 17th century had an average of 6 members compared to an average of 3.7 in the rest of the population.

15.3 *Urban space and relationships: the neighbourhoods in Matera in the 17th century*

Obviously, the family was not just a single entity, a small world centred around its own interests. A wide network of relationships, patronage and protection, friendship and solidarity, and credit and exchange revolved around it, intertwining with blood relationships and kinship.

A close-knit system of mutual support, but also one of conflicts and tensions, created solid relationships between people with the same jobs or who lived in the same courtyard, the same court or neighbourhood. A study of neighbourhood relationships in Matera in the 17th century provides some extremely interesting aspects. One of the oldest and most atmospheric cities in Italy, the urban structure of Matera is made up of a series of stone caves cut into the rocks, known as the *Sassi* (Sasso Baresano and Sasso Caveoso) which have been a world heritage site since 1993 and are under the auspices of UNESCO (Carbone 2010b).

The core of this human settlement in the *Sassi* were the neighbourhoods, small groups that extended beyond the family unit, in a sense, the origins of the old urban districts. The neighbourhoods of Matera were a united, organic unit both from a topographic and a social point of view, with an extremely unusual internal structure: the houses, caves and smaller grottoes, with walls of rock that were rarely whitewashed, usually only had one opening for air and light situated near the entrance which opened onto a communal area for all the families of the neighbourhood.

The neighbourhoods of Matera were undoubtedly shaped by the unusual morphology of the *Sassi*, and mainly housed poorer families. The buildings owned by the Church and the wealthy families were situated in the *Civita*, the highest point of the city and therefore far away from the sphere of the neighbourhoods.

In 1689 the parish register of San Pietro Caveoso recorded 3,076 inhabitants, divided into 760 families, with an average number of 4 members per family (Carbone 2005: 17-80). The parishioners were sub-divided into districts which were further divided into smaller neighbourhoods. Despite the undisputed and documented predominance of the nuclear family (70.1%), families were bound to each other by complex family and neighbourly relationships which were not always easy to identify. Forty-six small neighbourhoods are recorded in the 17th century document, with an average of 16 to 17 families in each one. However, the data do not reveal the significant differences in the population density of the neighbourhoods which were strictly correlated to the morphological structure of the area. There were, in fact, some neighbourhoods with only a few families (5 or 6 small families) and others with 30 to 40 households. Not without some difficulty, detailed study of the sources permitted identification of relationships between the families that lived in the same neighbourhood. The study of surnames, together with information on paternity and maternity of all male and female parishioners, proved to be a useful research methodology.

Despite the tendency to neolocalism and the indisputable prevalence of the nuclear family, the analysis of the relationships that bound families living in the same neighbourhoods highlighted the tendency of married children to choose their new home near to that of their parents. The choice of proximity to the family of origin does not seem to have been influenced by gender: married daughters or married sons could both find themselves living nearby. In other instances, married brothers and sisters lived in the same neighbourhood with their respective families.

Apart from family relationships, another factor that could create a bond within the close network of relationships between families of the same neighbourhood was the job of the head of the household.

The neighbourhood fulfilled the functions of mutual help, cooperation and social control, thus creating a close relationship between the life and history of the individual and that of the whole neighbourhood. Especially for the women, who were destined to caring for their children and homes, this provided the only opportunity for distraction, socialising and entertainment. In this way, the neighbourhood became a type of society of families, the fulcrum of community life in the two rocky districts of Matera.

However, while it is possible to attribute the neighbourhood with an institutional role and a specific social function, the negative aspects should not be overlooked: the close proximity of private and public space between some family units and the community could lead to conflicts, hypocrisy, social control, external influences and jealousies between people who were inevitably obliged to live their private lives in public due to a lack of living space resulting from the morphology of the land.

15.4 *Women and dowries: family strategies in the 18th and 19th centuries*

The traditional family in modern Italy was characterised by its extreme vulnerability to external influences and community control. Relationships between the different members of the nuclear family were deferential and authoritarian (Wiesner-Hanks 2017). Marriage was seen as a 'contract', mainly of an economic nature, for productive and reproductive purposes. It was a mechanism for passing on the family wealth from one generation to the next (Da Molin 2014: 149-181).

The *conditio sine qua non* to marriage was the dowry. The study of the marriage contracts related to some situations in Southern Italy permitted the reconstruction of family strategies and differentiated dynamics according to social-professional groups that placed the female figure at the centre of events.

From this information, social, economic and matrimonial agreements that were intended to protect and preserve professional identity emerged, for example, within highly skilled craft guilds registered in some areas of Apulia in the 18th century, such as the copper and ceramic crafts (Carbone 2000: 151-172; Da Molin and Carbone 2009: 305-324).[8] When there was no son to whom the workshop could be handed down, one of the daughters usually married a young man who practised the same craft as her father. In this way, the daughter became an indirect vehicle to guarantee the continuity of the craft, the financial inheritance and the handing down of the workshop to the descendants. A future husband was chosen from a restricted number of families who were bound by friendship and alliances between the main lineages, in the same neighbourhood or also from the same workshop as the father.

It is well-known that not all daughters were guaranteed a dowry and, consequently, the chance to marry. In many cases, women were obliged to respect the will of the family and were destined to taking forced monastic vows, or to spending their lives assisting in the household. This was a form of psychological violence, of control and coercion, which left no room for personal choice or freedom to marry. In other cases, the conflict between parents and children in the pre-nuptial period could determine the segregation of the woman in a female institute, with a fine line between assistance and reclusion.

An example of this was the case of Olimpia, whose father had died, and who was forcibly led by some of her relatives to the *Asilo di Pietà* (Institute of Mercy) in Bari in 1823 (Carbone 2010a: 13-37). Her paternal grandfather was in perennial conflict with his daughter-in-law whom he

---

8   A very important role was played by the daughters of specialized artisans. If the artisan did not have sons, a daughter would marry a young man who worked in the same craft sector. In this way, the workshop and the continuity of the family tradition were safeguarded.

reputed to be of low morals and a bad example to the girl. He did not condone the imminent marriage of the young woman to her betrothed, a tailor named Nicola, who he believed to be a philanderer.

In a society founded firmly on the safeguarding of honour, pre-marital virginity and conjugal loyalty, closely correlated to strong religious and institutional networks with a marked asymmetry of the male-female roles conditioned by rigid stereotypes, the inappropriate behaviour of a daughter could lead to her expulsion from the family.

A pre-marital or extra-marital pregnancy, unruly behaviour or the non-acceptance of the shared rules meant a woman was destined for the fringes of society and permanent social stigma.

15.5 Conflicts and legal papers: violence against women in and outside the family in the 19th century

19th century trial documents from the Crown Court in Bari were studied in order to further our knowledge on the complex world of family relationships. Conflictual relationships between spouses, parents and children and relatives often culminated in deviant and transgressive behaviour (Da Molin, Carbone and Loprieno 2011: 33-50).

The reconstruction of violence against women based on the trial reports is only the tip of the iceberg and does not claim to offer a plausible picture of the reality, a reality reluctant to be classified, and provides the scholar with only a partial outline from the official documents. In fact, many acts of violence, especially those involving young women, remained hidden, wisely concealed within the domestic walls, protected by the silence of the victims who were mostly concerned with defending their honour. Shame, fear of having their future permanently tainted, but also bonds of affection, and psychological and material dependence between the victim and her torturer were often welded together in a ring of conspiratorial silence and reticence.

Crimes such as steering others into prostitution, rape, incest and uxoricide fill the dense trial reports that offer significant glimpses of ordinary life and aspects of daily family life.

In 1880, Angela, «a woman of low morals and depraved habits» from Gravina was accused of prostituting her 12-year-old daughter, Maria.

In January 1890, the entire town of Bitonto, in the province of Bari, was shocked by rumours: Rosina, a 15-year-old girl, had been raped by her father Giuseppe, a 42-year-old baker, and was pregnant. Scandal and indignation upset the peace and quiet for months until, confronted with the evidence of the girl's pregnancy, the man was condemned to 10 years in prison and deprived of his parental responsibility as he was «guilty of carnal knowledge with violence and abuse of power over the person of his daughter Rosa».

Domestic violence against women reaches the height of cruelty in uxoricide, a transversal phenomenon that knows no geographical boundaries, cultural distinctions, social status or age, both today and in the past.

Poverty and constant cruelty culminated in the murder of 26-year-old Lucrezia by her husband Stefano, a 29-year-old shoemaker, on the evening of 5th January 1870 in Putignano (Bari). Six stab wounds, two of them fatal, killed the young woman. The husband's excuses during the proceedings were worthless: all the witnesses testified that Lucrezia was «honest and hard-working' and the accused was a 'good-for-nothing». Declared guilty of the willful homicide of his wife, Sebastian was sentenced to a life of forced labour which was then converted to 24 years' imprisonment and 2 years' special surveillance.

However, danger also lurked outside the domestic walls.

In a social context where private and public life constantly intermingled, the conditions that young girls lived in, often with no family control, made it easy for people with bad intentions to approach them. Youth, physical weakness and sexual ignorance all made childhood fertile terrain for sexual predators.

Many young girls were not able to put up any resistance to their attackers: they did not understand the underlying signs and when they realised the danger, by instinct or pain, it was too late.

On 29th December 1863, Maria was only four years old when the man «with a scar on his cheek», whom she knew well as he lived near her father's home, invited her into a nearby windmill where he brutally raped her. The accused's previous criminal record aggravated his position, as he had already been condemned for the violent rape of a 6-year-old girl.

In other cases, the hidden danger lay inside the master's workshop where, from a young age, girls were sent to learn a trade. This was the case of 16-year-old Antonia who was raped by her master, a tailor, and became pregnant. She was forced to abandon her newborn daughter, the fruit of violence and dishonour, in the foundling wheel at the entrance to the home for foundlings, where the baby died the same day.

Bibliographical references

Ago, R., and B. Borello (eds), *Famiglie. Circolazione di beni, circuiti di affetti in età moderna* (Rome: Viella, 2008).

Alfani, G., *Padri, padrini, patroni. La parentela spirituale nella storia* (Venice: Marsilio, 2006).

Alfani, G., and V. Gourdon, 'Entrepreneurs, Formalization of Social Ties, and Trustbuilding in Europe (Fourteenth to Twentieth centuries)', *Economic History Review*, 65, 3 (2012a), 1005-1028.

— *Spiritual Kinship in Europe, 1500-1900* (Basingstoke: Palgrave Macmillian, 2012b).

Anderson, M., *Family Structure in Nineteenth Century Lancashire* (Cambridge: Cambridge University Press, 1971).

— *Interpretazioni storiche della famiglia. L'Europa occidentale 1500-1914* (Turin: Rosenberg & Sellier, 1982).

Arru, A., 'Servi e serve: le particolarità del caso italiano', in *Storia della famiglia italiana, 1750-1950*, ed. by M. Barbagli and D. I. Kertzer (Bologna: il Mulino, 1992), pp. 273-306.

— *Il servo. Storia di una carriera nel Settecento* (Bologna: il Mulino, 1995).

Barbagli, M., *Sotto lo stesso tetto. Mutamenti della famiglia in Italia dal XV al XX secolo* (Bologna: il Mulino, 2013).

Berkner, L. K., 'The Stem Family and the Development Cycle of the Peasant Household', *American Historical Review*, LXXVII (1972), 398-418.

— 'The Use and Misuse of Census Data for the Historical Analysis of Family Structure', *Journal of Interdisciplinary History*, V (1975), 721-738.

Bernardi, U. (ed.), *Famiglie e sviluppo sociale nelle opere di Frédéric Le Play* (Milan: Jaca Book, 1981).

Caporrella, V., *La famiglia. Un'istituzione che cambia* (Bologna: Archetipolibri, 2008).

Carbone, A., 'La via del rame. Mestieri, strategie matrimoniali e sistemi dotali in Terra di Bari a metà Settecento', *MEFRIM, Mélanges de l'École française de Rome*, tome 112, 1 (2000), 151-172.

— *Vita nei Sassi. Famiglia, infanzia e assistenza a Matera in età moderna* (Bari: Cacucci Editore, 2005).

— *Dall'Albergo delle Pentite all'Asilo di Pietà: reclusione e assistenza alle donne a Bari fra Settecento e Ottocento*, in *Orizzonti di ricerca*, ed. by Dipartimento di Scienze Storiche e Geografiche (Bari: Cacucci Editore, 2010a), pp. 13-37.

— *Tra vicoli e precipizi. Popolazione, società e istituzioni a Matera nel corso del Settecento* (Bari: Cacucci Editore, 2010b).
Casanova, C., *La famiglia italiana in età moderna. Ricerche e modelli* (Rome: Carocci, 1997).
— *Famiglia e parentela nell'età moderna* (Rome: Carocci, 2009).
Clegg, J., 'Good to Think with: Domestic Servants, England 1660-1750', *Journal of Early Modern Studies*, 4 (2015), 43-66.
Da Molin, G., 'Struttura della famiglia e personale di servizio nell'Italia meridionale', in *Storia della famiglia italiana, 1750-1950*, ed. by M. Barbagli and D. I. Kertzer (Bologna: il Mulino, 1992), pp. 219-252.
— *Famiglia e matrimonio nell'Italia del Seicento* (Bari: Cacucci Editore, 2000).
— *Storia sociale dell'Italia moderna* (Brescia: Editrice La Scuola, 2014).
Da Molin, G., Carbone, A., and M. Loprieno, 'Fanciulle violate: i processi criminali a Bari nel XIX secolo', in *Ritratti di famiglia e infanzia. Modelli differenziali nella società del passato*, ed. by G. Da Molin (Bari: Cacucci Editore, 2011), pp. 33-50.
Da Molin, G., and A. Carbone, *Carte d'archivio. Storia della popolazione italiana tra XV e XX secolo* (Bari: Cacucci Editore, 2016).
— 'Gli artigiani nel Mezzogiorno d'Italia nel XVIII secolo: modelli differenziali della famiglia, del matrimonio e del controllo degli assetti produttivi', in Fondazione Istituto Internazionale di Storia Economica "F. Datini" Prato, *La famiglia nell'economia europea (Secc. XIII-XVIII)*, Atti della "Quarantesima Settimana di Studi", Prato 6-10 aprile 2008, ed. by S. Cavaciocchi (Florence: Firenze University Press, 2009), pp. 305-324.
Delille, G., *Famille et propriété dans le Royaume de Naples (XVe-XIXe siècle)* (Rome-Paris: Ècole française de Rome – Édition de l'École des Hautes Études en Sciences Sociales, 1985); It. trans. *Famiglia e proprietà nel Regno di Napoli, XV-XIX secolo* (Turin: Einaudi, 1988).
— *Famiglia e potere locale. Una prospettiva mediterranea* (Santo Spirito, Bari: Edipuglia, 2011).
Durkheim, É., *Les règles de la méthode sociologique* (Paris: Alcan, 1910).
— *De la division du travail social* (Paris: Alcan, 1911).
— *La sociologie* (Paris: Larousse, 1915).
Fazio, I., and D. Lombardi (eds), *Generazioni. Legami di parentela tra passato e presente* (Rome: Viella, 2006).
Guttormsson, L., and P. P. Viazzo, 'Il servizio come istituzione sociale in Islanda e nei paesi nordici', *Quaderni Storici*, 68 (1988), 355-379.

Hajnal, J., 'Age at Marriage and Proportion Marrying', *Population Studies*, VII, n. 2 (1953), 111-136.

— 'European Marriage Patterns in Perspective', in *Population in History*, ed. by D. V. Glass and D. E. C. Eversley (London: Arnold, 1965); It. trans. *Modelli europei di matrimonio in prospettiva*, in *Famiglia e mutamento sociale*, ed. by M. Barbagli (Bologna: il Mulino 1977), pp. 267-312.

— *Two Kinds of Pre-Industrial Household Formation System*, in *Family Forms in Historic Europe*, ed. by R. Wall, J. Robin and P. Laslett (Cambridge: Cambridge University Press, 1983), pp. 65-104; It. trans. 'Due tipi di sistemi di formazione dell'aggregato domestico preindustriale', in *Forme di famiglia nella storia europea* (Bologna: il Mulino, 1984), pp. 99-142.

Hareven, T. K., and A. Plakans (eds), *Family History at the Crossroads. A Journal of Family History Reader* (Princeton: Princeton University Press, 1987).

Herlihy, D., and C. Klapisch-Zuber, *Les Toscans et leurs familles: une étude du catasto florentin de 1427* (Paris: Presses de la Fondation Nationale des Sciences Politiques, 1978); It. trans. *I toscani e le loro famiglie. Uno studio sul catasto fiorentino del 1427* (Bologna: il Mulino, 1988).

Hinde, P. R. A., Ago, R., and A. Torrente, 'L'influenza del servizio rurale e domestico sulla demografia inglese, 1850-1914', *Quaderni Storici*, 68 (1988), 541-571.

Kertzer, D. I., and R. P. Saller (eds), *The Family in Italy from Antiquity to the Present* (New Haven: Yale University Press, 1991).

Lanzinger, M., and R. Sarti (eds), *Nubili e celibi tra scelta e costrizione (secoli XVI-XX)* (Udine: Forum, 2006).

Laslett, P., 'La famille et le ménage', *Annales E.S.C.* (1972), 847-872; It. trans. 'Famiglia e aggregato domestico', in *Famiglia e mutamento sociale*, ed. by M. Barbagli (Bologna: il Mulino, 1977), pp. 30-54.

Le Play, P. G. F., *La réforme sociale en France*, 2 vols (Paris: Henri Plon, 1864).

Levi, G., *Centro e periferia di uno stato assoluto. Tre saggi su Piemonte e Liguria in età moderna* (Turin: Rosenberg & Sellier, 1985a).

— *L'eredità immateriale. Carriera di un esorcista nel Piemonte del Seicento* (Turin: Einaudi, 1985b).

Livi, L., *La composizione delle famiglie: studio demografico* (Florence: Tipografia M. Ricci, 1915).

Raffaele, S., *Famiglie e senza famiglia. Strutture familiari e dinamiche sociali nella Sicilia moderna* (Naples: Edizioni Scientifiche Italiane, 2000).

Reher, D. S., 'Family ties in Western Europe: Persistent Contrasts', *Population and Development Review*, 24 (1998), 203-234.

Saraceno, C., and M. Naldini, *Sociologia della famiglia*, 3rd edn (Bologna: il Mulino, 2013).

Sarti, R., 'Who are Servants? Defining Domestic Service in Western Europe (16th-21st Centuries)', in *Proceedings of the "Servant Project"*, ed. by S. Pasleau and I. Schopp with R. Sarti, 5 vols (Liège: Éditions de l'Univérsité de Liège, 2005, but 2006a), vol. 2, pp. 3-59.

— 'Domestic Service: Past and Present in Southern and Northern Europe', *Gender & History*, 18, n. 2 (2006b), 222-245.

— *Vita di casa. Abitare, mangiare, vestire nell'Europa moderna*, 4th edn (Rome-Bari: Laterza, 2017).

Shorter, E., *The Making of the Modern Family* (London: Collins, 1976).

Sovič, S., Thane, P., and P. P. Viazzo (eds), *The History of Families and Households: Comparative European Dimensions* (Leiden, The Netherlands: Koninklijke Brill NV, 2016).

Stone, L., *Family, Sex and Marriage in England, 1500-1800* (New York: Harper & Row, 1979).

— *Viaggio nella storia* (Rome-Bari: Laterza, 1987).

Tamassia, N., *La famiglia italiana nei secoli decimoquinto e decimosesto* (Milan: R. Sandron, 1911).

Wall, R., Hareven, T. K., and J. Ehmer (eds), *Family History Revisited: Comparative Perspectives* (Newark: University of Delaware Press, 2001).

Wiesner-Hanks, M. E., *Le donne nell'Europa moderna* (Turin: Einaudi, 2017).

Wrigley, E. A., and S. R. Schofield, *The Population History of England 1541-1871: A Reconstruction* (London: Arnold, 1981).

GIOVANNA SUMMERFIELD

16.
INTERDISCIPLINARY CLASSROOMS AS INNOVATIVE SPACES TO RESEARCH AND DISCUSS CONFLICTS AND POSSIBLE CHANGES

As «history and sociology are both concerned with human social practice in its capacity for... change and also in its tendency to reproduce itself» (Steinmetz 2007:1), an interconnectivity between the two disciplines is indispensable for U.S. university learners while addressing global issues through research, discussion and action. This need assessment is not new in the minds of intellectuals; indeed, already in the middle of the nineteenth century, Auguste Comte expressed his concerns about the compartmentalization of human knowledge, so endemic to our educational institutions. According to him, while specialization allows for a «felicitous development of the spirit of detail... [it] spontaneously tends... to snuff out the spirit of togetherness, or at least to undermine it profoundly» (Kapp 1961:60). In order to face this challenge, I have adopted a transdisciplinary[1] approach in the context of my instruction, not only attempting, in an integrated manner, to address issues that cross disciplinary boundaries, but also prioritizing active collaboration with the individuals affected by the issues and their exploration, and community-based stakeholders, as opposed to a more interdisciplinary style of instruction and research. This transdisciplinary approach is particularly valuable when focusing on

---

[1] "Whereas interdisciplinary programs start with the discipline, transdisciplinary programs start with the issue or problem and, through the process of problem solving, bring to bear the knowledge of those disciplines that contributes to a solution or resolution." Meeth writes (as recorded on p. 6 of *Beyond the Boundaries: A Transdisciplinary Approach to Learning and Teaching* edited by Douglas Kaufman, David M. Moss, and Terry A. Osborn, 2003). Whereas multidisciplinary keep the different disciplinary voices separate and the interdisciplinary integrate those voices, the transdisciplinary collaborations transcend the academic research including external stakeholders as well (David BudtzPedersen, 'Integrating Social Sciences and Humanities in interdisciplinary Research', *Palgrave Communications* 2, 5 July 2016, p. 14).

international relations and interdependence, and when awareness regarding global issues needs to be raised. It is imperative to keep in mind that

> Today social studies should include a world-centered treatment of humankind, [taking into account] the variety of actors on the world stage... and the linkages between past actions, present social, political, and ecological realities and alternative futures where students should perceive the close relationships between past, present and future....The curriculum should demonstrate that individuals and groups can influence and can be influenced by world events. Furthermore, social studies curriculum should help develop the understanding, skills, and attitudes needed to respond effectively and responsibly to world events (NCSS 2016).

As educators, we should want our students to be better equipped with a number of intellectual and practical competencies as well as crosscutting skills that will allow them to analyze and potentially solve issues from different perspectives, with a method that implies not only open communication with various key agents, but an approach that aims to comprehend the complexity of problems and stakeholders and the connections between them. Many scholars and decision makers today cannot deny that most of the issues we face do require integration of diverse knowledge and approaches (Pedersen 2016: 3).With an interdisciplinary background in political science, romance languages and literatures, European and Mediterranean history, women's studies as well as global leadership and civic engagement, I am personally one of those educators who does not need much convincing or training to conduct a course of study and research that is holistic in nature.

In this essay, I will address the objectives, perspectives, and some of the practical activities used in two courses I taught in recent years, more precisely an advanced undergraduate course titled "East Meets West: Sicily, a Case Study at the Crossroads", during the spring semester of 2014, and an Honors class titled "Global Citizenship", during the spring semester of 2016, both at my home institution. Upon completion of the first course, the students were to: be familiar with and understand the cultures of the Mediterranean, including and underscoring the culture of Sicily; recognize the socio-economic impact of the Mediterranean and Sicily on national and international societies; develop identities as community members and recognize the power of individual and collective work; demonstrate ability to work in pairs and teams within and outside campus; critical thinking, communication, and research skills. The objectives of the second course were the following: to develop a broader understanding of issues of key concern to the international community, and a broader understanding and appreciation of diverse cultures while being familiar with the concepts of citizenship and global citizenship, cosmopolitanism, transnationalism, and multiculturalism to name just a few; and to develop the ability to investigate, consider, and debate avenues which may be open in order to participate as active citizens within their social, cultural, and professional contexts. Regardless of the topics and the student body of the two courses, history mattered

> because without a credible story about where we have been before, we truly have no idea where we are now. And without evidence about past sequences of cause and effect, it is well-nigh impossible to develop intelligent plans for the future (McCants 2016).

Regardless of the topics and the student body of the two courses, a sociological perspective was also applied to identify «patterns in human interaction, how and why these patterns exist, the consequences of them, and how to reproduce or change the patterns» (Steele 2008: 4).

Moreover, regardless of the topics and student body of the two courses, the learners were engaged in producing persuasive documents or media presentations, demonstrating the capacity to apply their skills to try and solve thorny issues while acquiring a well-developed sense of empathy, understanding, and passion. They were to communicate also through social media to continue and expand conversations with peers at local and international levels. Finally, both courses involved research contributions to conferences: for the first course, a Mediterranean Studies international conference held in June 2014 at the University of Catania, Sicily, cosponsored with the department of political sciences and other academic units of the same institution; and for the second course, a diversity symposium held at the Center for the Arts and Humanities of the College of Liberal Arts at Auburn University in March 2016, coincidentally taking place on International Women's Day, March 8.

In the first course, "East Meets West: Sicily, a Case Study at the Crossroads", the island was presented as a case study to help students be more informed about the innumerable civilizations that have invaded and controlled Sicily over three thousand years (i.e. Phoenicians, Greeks, Romans, Byzantines, Arabs, Normans, Spaniards, and Bourbons). It was treated as a window into the past to face our multicultural present and imminent future; it provided the groundwork to learn, analyze, discuss, and apply the information presented on different fronts with a stimulating multifaceted approach, which was geographical, historical, sociological, literary, political, and religious. The course was divided into two parts: the first part addressed issues of identity, covering the temporal and spatial contexts, while the second part examined the structures and legacies that connect the Mediterranean cultures. The class included a series of lectures, readings, and film screenings, as well as a team of national and international leading experts in Mediterranean studies, world history, comparative literature, cultural geography, and art history. It was a hybrid course that drew on such concepts as active learning and student engagement, while utilizing a "flipped classroom design," meaning that during class sessions the instructor functioned mainly as a coach or advisor, encouraging the students involved in individual inquiries and collaborative efforts, and stimulating discussions that occurred in and outside the classroom. Students were given iPads loaded with apps that would foster civic engagement and individual research on topics of special interest within the general context of the course, mainly multiculturalism, cultural pluralism, assimilation, acculturation, ethnicity, religion, gender, class, and migration.

For his/her mid-term project, each student had to represent one specific civilization and its impact and join a group with the same interest. The project was called "Which side are you on?" and students were required to abide by the following guidelines: "adopt" a civilization that affected Sicily; describe the reason for their selection; describe the reasons why this civilization was interested in Sicily; its success in controlling Sicily; the way its goals were carried out and the final outcomes; explain whether theirs was an assimilation or integration; and narrate one or two events or characteristics of this civilization that was really impressive or surprising and/or that were influential, (un)just, (in)effective, and the impact they had on Sicily. Students were also supposed to report their self-assessment, through role playing (for example, if they were living in Sicily at that time as locals, how would they have perceived this civilization? If instead they belonged to the civilization coming into Sicily, how would they have perceived Sicily? Finally, with the hypothetical possibility of going back in time, list things that they would *and* would not do, both as Sicilians and as "guests"), and through research questions (stances, conclusions they had drawn out of their study, applications of their study to contemporary situations, with an extemporaneous comparison with the audience) were the other students convinced of the "pros and cons" of the civilization discussed?

The groups presenting had to answer this series of questions using a Blog, iBook, Power Point, video, or other digital media of their choice (if using PowerPoint, Prezi or video, students needed to turn in an additional written report with answers to all the above questions, which was not required for the other formats). Whereas creativity was encouraged, an extensive and appropriate bibliography was mandatory for all formats selected. Students were also told that their projects, after scheduled class presentations, timed at 12 minutes with 5-8 minutes of discussion, were to become public and shared with their Italian counterparts and their leading faculty. For grading purposes, they had clear assessment criteria for content and delivery handed prior to presentation and after, with written evaluation.

The final project comprised an individual bibliography and plan of study to work on a research assignment with faculty approval. Particular inspiration came from discussions on issues analyzed in the latter part of the courses, which entailed a clear appreciation of one's own cultural and socio-political setting as compared with the Mediterranean one. Some of these topics were honor, hospitality, immigration, and emigration. Students were able to connect with our international partner, the Euro-Mediterranean Center in Catania, and other local graduate students. Gone was the sense of uniqueness and remoteness students had felt at the beginning of the course; instead, a commonality of interests and resources that could benefit the two different contexts and diverse populations was perceived. In their reflections, students were aided by the skits performed by the Mosaic Theatre Company, an Auburn student theatrical ensemble whose primary purpose is to facilitate dialogue about diversity through performance.

The second course, "Global Citizenship", was a predominantly discussion-based seminar, coupled with a reading list and a plethora of guests, activities, and online resources, which offered a foundation in historical and contemporary concepts and the challenges of global citizenship. We pondered questions like "What does it mean to be a global citizen?" "Who is a citizen?" "What does 'citizenship' imply?" "Is it possible to be a global citizen?" while examining key challenges to global citizenship (the challenges of being informed in face of media influence, and cultural and social diversities) and considering a number of fundamental issues that are of great significance to the international community. Students were asked to consider these concepts and issues from their own experiences, life skills, and disciplinary knowledge, to reflect on their responsibilities and to assume positions of advocacy in areas of interest on a local level to which they felt they could most contribute and for which they felt most concerned and passionate. Students were also encouraged to undertake further research in their areas of choice and to share useful resources, including current newspapers headlines, with their peers in the classroom. Since the course was divided into three modules (1. Concepts from historical and modern perspectives; 2. Citizen-led movements in different geographical areas and times; 3. Pressing issues and possible actions), students travelled virtually in order to obtain the assistance of international speakers and field experts around the world to explore global policies and phenomena. This course was taught in one of the facilities we call EASL – Engaged in Active Student Learning – classrooms. The new EASL classroom at Auburn University is a student-centered space featuring table clusters that accommodates six students each. Each cluster has its own glass board for writing and sharing ideas, as well as a monitor so students can connect their electronic devices and make use of web-based resources during the learning process. The EASL classrooms are often referred to as "flipped classrooms" due to a non-traditional, or inverse teaching method that is applied in the space. Instead of lecturing in class, a faculty member might record the lecture on a video and make it available on the Internet before the class meets. Then students will apply their learning on their own in groups in the classroom. The professor may give some examples or suggest resources before class; then in the classroom the students do activities, conduct experiments, and/ or have discussions, so they can apply the knowledge in the classroom and the faculty member can gauge the students' performance. Through this pedagogy, students learn how to communicate and how to become leaders.

At the beginning of the semester, the students became familiar with the concept of citizenship through major thinkers like Diogenes, Epictetus, Aristotle, Plato, Cicero, Machiavelli, Rousseau, to name a few. Students also attended a special lecture by Arun Gandhi, grandson of the legendary Mohandas Gandhi, titled "Lessons Learned from My Grandfather: Nonviolence in a Violent World." On that premise, students started to learn about global movements in response to problematic political or cultural situations, such as the independence of the United States, South Africa Apartheid, Tiananmen Square student protest, Zapatista revolt, women in the MENA and Arab Spring in general, and Addiopizzo, aided by guest speakers and docu-film screenings. Every guest speaker was a specialist in a different field: political science, Asian Studies, sociology, history, art history and Italian Studies. These analyses and discussions helped them to draft their research and presentations. The latter part of the course was focused on current pressing issues, as assessed by the Gallup Polls in the States and Europe. Each student worked within a group of choice to present the facts; they looked directly at the causes and consequences and proposed some solutions, after watching some helpful videos, such as the Universal Declaration of Human Rights' public service announcements and the TED Talks Copenhagen Consensus delivered by Bjorn Lomborg and Simon Anholt'sWhich Country. The course ended with a clear understanding of the individual responsibilities and impact of/on their communities and of potential strategies, i.e. Marshall Ganz's storytelling and mobilizing communities to include the power of social media and different other venues that we analyzed and used throughout the duration of the course.

In both courses, the students recognized that the world is interconnected and interdependent, and that although they might have had a matrix in the past, the ways in which the world is currently interconnected, are ever changing at present, thus creating new outcomes. Students understood that global issues impact on local contexts, thus recognizing the importance of being aware of world affairs and consequences. They could critically analyze the ways that global interconnectedness contributes to iniquities within and beyond individual nations; most importantly, they understood and wanted to find ways they could pursue to address some of these issues at the local, national or international levels.

In both settings, students started working progressively at an individual level to acquire more familiarity with complex issues while researching on their own and then sharing with their peers, going from a debate format with two sides opposing each other as in the mid-term project, seeking to affirm their own point of view and assumption and defending their own position as the best solution, to a deliberative dialogue format, which involves two or more sides working collaboratively toward common understanding, since deliberative dialogue participants listen to other perspectives in order to understand, find meaning, and reach agreement. With finding common ground as their goal, the students attempted to keep an open mind, and reevaluate, weigh, expand, and possibly change their own points of view (Gunzman 1999). Since a deliberative dialogue is less confrontational than a debate, it offers a way to discuss crucial issues and possible strategies. It is reflective and requires people to listen to one another, without trying to persuade, but to understand key interests and values that will help determine what action is most suitable and effective for all parties concerned. With the assistance of the moderator (not necessarily an expert), in this case the faculty, the group's goal is to identify the common ground and the group's shared interests, clarify agreements, disagreements and trade-offs, and conclude with attainable negotiations to determine a plan of action (Buchanan 2001: 1-2). In order to help build leaders who think critically, reflect on actions, and compromise in the face of conflict, teachers must intentionally connect learning in the classroom to the world outside the classroom. High-tech tools and interactive programs enable the "sage on the stage" to be transformed into the "guide on the side" (Morgan 2016: A12), with the students as main stakeholders and partners.

Nelson Mandela once stated that education is the greatest avenue for changing the world. He asserted that educating our youth plays a critical role in spurring social, economic and political progress (2003). Through an innovative pedagogical methodology with collaboration involving not only the students themselves but also teachers and scholars with diverse backgrounds and intellectual interests and expertise, and especially through the realization that individual and collective actions can be taken at all levels to make changes, we were able to facilitate not only academic ownership, understanding the interconnectivity and importance of academic disciplines, but also personal empowerment and community partnerships. The students, in the courses described here and offered in the College of Liberal Arts at Auburn University, were able, through a transdisciplinary approach, to link different fields of study, different geographical and cultural areas, and different perspectives to posit and apply strategies for the common good.

Bibliographical references

Anholt, S., *Which Country Does the Most Good for the World?*, Berlin: TEDSalon, 2014, https://www.ted.com/talks/simon_anholt_which_country_does_the_most_good_for_the_world?language=en [accessed 5 February 2016]

Buchanan, A., *Inclusion and Diversity: Finding Common Ground for Organizational Action. A Deliberative Guide* (Ottawa, Canada: Canadian Council for International Cooperation, 2001).

Guzman, J., 'What is Deliberative Dialogue?', Insights on Educational Policy, Practice, and Research, 9 (October 1999) http://www.sedl.org/policy/insights/n09/1.html. [accessed 10 November, 2016.]

Kaufman, D., Moss, D. M., and T. A. Osborn (eds), *Beyond the Boundaries: A Transdisciplinary Approach to Learning and Teaching* (New York: Praegers Publishers, 2003).

Kapp, W., *Toward a Science of Man in Society: A Positive Approach to the Integration of Social Knowledge* (The Hague, Netherlands: Martinus Nijhoff, 1961).

Lomborg, B., *Global Priorities Bigger than Climate Change*, TED2005, http://www.ted.com/talks/bjorn_lomborg_sets_global_priorities?language=en [accessed 17 January 2016]

Mandela, N., 'Lighting the Way to a Better Future', Address at Launch of Mindset Network, (Johannesburg, 16 July 2003) http://www.mandela.gov.za/mandela_speeches/2003/030716_mindset.htm [accessed 20 November 2016].

Mangan, K., 'How One University Encourages Innovation in Teaching', *Chronicle of Higher Education*, LXIII (16), (9 December 2016), A12-13.

McCants, A., 'How History Can Help Us Solve Global Economic Issues.' *MIT News,* 22 April 2016.

National Council for the Social Studies Board of Directors, 'Global and International Education in Social Studies' https://www.socialstudies.org/positions/global_and_international_education [accessed 10 November 2016]

Pedersen, D. B, 'Integrating Social Sciences and Humanities in interdisciplinary Research', *Palgrave Communications*, 5 July 2016, 1-28.

Steele, S. F., and J. Price. *Applied Sociology: Terms, Topics, Tools and Tasks* (Belmont, CA: Thomson Higher Education, 2008).

Steinmetz, G., 'The Relations between Sociology and History in the United States: The Current State of Affairs', *Journal of Historical Sociology,* 20 (1-2) (March-June 2007), 1- 12.

'United for Human Rights', http://usahumanrights.com/30-public-service-announcements [accessed 10 February 2016]

Anna Maria Leonora

17.
Soft Revolution! Education Systems in Conflict: Standard Versus Non-Standard Forms of Socialization in a Gender Perspective

17.1 *The Theoretical Background of the Research*

At the end of the 1990s, the most authoritative supranational bodies expressed both perplexity and hope with regard to the possibility of preparing properly for the changes taking place in the lives of individuals and in the global social context from a perspective considered to be effective: the educational process. On the one hand, already twenty years ago the report *Unesco Learning: the treasure within* (Delors 1996) showed indications that educational processes had changed to such an extent by the challenge of multiculturalism and globalization and the consequent structural crisis that there was a 'before' and 'after' divide, that was beginning to distort the geopolitical landscape which had been taken for granted until then (Kymlicka 1998). On the other hand, the White Paper of the European Union *Insegnare e apprendere. Verso la società conoscitiva –* Teaching and Learning. Towards the knowledge society (COE 1995) clearly acknowledged the awareness that the known and tested models of training processes were partly outdated and unable to respond to the change that was already taking place in the societies of the democratic West; this was also the case in several other different contexts regarding culture, economics, and politics. Despite everything, education and training processes were and are still considered today to be the most important and promising strategies to ensure a progressive, positive and democratic direction of global society (IEP 2020).

It is clear that today these idealistic intentions have been largely disregarded, giving way to other priorities and emergencies perceived as such by the policies of the nation states. In particular, the Italian school still oscillates between an organization dedicated to training individuals to be useful for economic growth, and a social institution focused on the moral maturity of people and citizens which is necessary to ensure social cohesion.

In concrete terms, schools, especially state schools, have lost their central role as learning contexts in the last twenty years, giving way to other areas and experiences that stimulate or support the quality of learning itself (Besozzi 2017; Bottani 2013; Di Pol 2016).

The Italian school has also continued to change at an increasingly fast pace in the international scenario, finding itself in an increasingly precarious situation which, however, it has tried to overcome by introducing new educational paths and models, and new disciplines together with new relational practices with those involved in the educational process: families, students and administrative institutions. This has not avoided the continuous deterioration of vertical and horizontal relationships (Colombo 2017) typical of the educational context, as is unfortunately also revealed by the growing number of school dropouts in the different levels of the educational path (Istat 2019). This data is often accompanied by further critical points: according to the surveys carried out with the INVALSI tests (2019), the results in relation to students' basic skills are lower overall in the South and the Islands and show significant differences between one school and another and even between one class and another.

Moreover, Italy has the highest percentage of teachers over the age of 50 among OECD countries (59%) together with the lowest percentage of teachers between 25 and 34 years old (OECD 2019).

At this point, it is important to underline the fact that the recognition of the crisis of the standard education model calls for greater consideration of non-institutional and non-standard education practices; these are organized according to alternative paths of a different nature and ideological reference. There are many possible variations of alternative practices and forms of non-standard education (see table 1[1]). In a very general perspective, the analytical reflection that follows is directed towards social contexts that reveal the link between the paths of socialization/education and the processes of social change read through an analytical prism that represents the non-standard educational choice *par excellence*: homeschooling. This approach to the education and socialization of children emphasizes how different family life projects define different and alternative processes of education feeding social change towards new forms of social relationships.

Anglo-Saxon literature (Greenwalt 2019; Kunzman and Gaither 2013) has developed a specific and systematic corpus on this issue that examines numerous themes (Gamuzza 2013), including parental motivations (Murphy 2012), academic achievement (Snyder 2013), the comparison

---

1 The table is an updated re-edition of the same published in the essay Leonora A. M. (2019), Seeing a tree grow. The upside-down perspective of non-standard educational processes in Italy, in: "Annals of the Faculty of Education Sciences", University of Catania, 18, p. 137.

of forms of legislation on home schooling (Valero 2013), the historical evolution of homeschooling practices (Gaither 2009; 2017), the role of mother in leading homeschooling practices (Jolly at alii 2013), just to mention a few of the in-depth perspectives.

> A substantial literature on the issue...but from those countries with a long tradition on homeschooling practices (UK; USA) "[...] homeschoolers are a notoriously difficult demographic to study because of the diversity of individuals engaged in the practice, the deinstitutionalized nature of the phenomenon, and the distrust with which many homeschoolers regard external surveillance" (Kuzman and Gaither 2013: 5).

Another interesting proposal is the one made by R. Morton (2010: 50) who identifies three types of homeschoolers in the UK, focusing on the decision-makers: parents who make a natural choice (looking for a healthier lifestyle than the industrial society), parents who make a social choice (to protect their children from environments that impose unwelcome values), and parents who choose the practice of homeschooling because it represents the only form of education that can be adapted to an unconventional lifestyle (for example, families of artists or people involved in dance, theatre, music or the circus).

This work is linked to a specific research experience, "*Homeschooling in Italia. La dimensione informale e collettiva delle pratiche educative* – Homeschooling in Italy. The informal and collective dimension of educational practices", which is part of the broader project FIR 2014-2016 "Ambiguities and conflict in spaces of belonging and sociality from the modern to the current" of the University of Catania. With regard to homeschooling and the particular subject of the above mentioned research, this essay performs a meta-analysis on ethnographic data collected and interpreted in a gender perspective, thus leading to new typological considerations regarding the non-standard education phenomenon already introduced and addressed in previous essays that are part of the theoretical premise of this work (Leonora 2014; 2019).

17.2 *A reasonable choice: the ethnographic approach*

Starting from the first contact with the world of homeschooler families, the Italian phenomenon of homeschooling, although limited in number and geographical spread (MIUR 2018), has progressively required more and more "sociological" attention; this characteristic of niche phenomenon concerning social change in training processes has suggested a research path that has taken on an exploratory dimension related to the forms of promotion and support of the homeschooling phenomenon as a non-standard form of education – an alternative to the path of state education.

The literature available in Italy in recent years (Chistolini 2009; Codello and Stella 2011) has favored the formulation of ideas and research questions on a phenomenon that, albeit marginal, is part of the sociological dimension of change as a conflict between two social parties. Given the small number of the social units that carry out homeschooling in Italy and their elusiveness, the most reliable source of information on the phenomenon remains contact with the people involved. The opportunity to interview them presented a considerable obstacle in the need to get in touch with a gatekeeper who would allow access to the field of investigation guaranteeing reliability and a non prejudicial view of the vital world of homeschooling.

We therefore decided to follow the insider perspective – according to the classic orientation of the Chicago School (Silverman 2000; Wax 1971) – making the researcher an interactive part in the investigated phenomenon that meant physical participation in the experience of the social units under analysis. In this way, it was possible to access not only the field of investigation, thus conducting a truly informative and experiential survey but – even more – to access the meaning of the practices and objectives of the social units involved. The proximity to the subjects involved in the homeschooling scenario led to the direct experimentation of educational practices based on non-standard learning models (e.g. the libertarian approach or other mixed methods inspired by different experiences such as Sumner Hill). Moreover, the decision to interpret the data from an insider perspective led to a grounded analytical strategy (Strati 1997; Charmaz 2006) that allowed us to synthesize the distinctive features of the different educational processes (see table 1) and to inductively outline three original typological figures with regard to the examined phenomenon (see table 2).

Table 1 - Synoptic and comparative framework Teaching Model (adaptation from Leonora 2019)

Teaching Model Analytical dimensions	Unschooling	Home-schooling	Libertarian Education	Outdoor School	Standard School
Definition	Literally non-school. This approach avoids any structural-impositional reference.	The educational path is chosen, managed, and implemented by the parents and mainly within the familiar context (spatial and relational).	Learning path based on the principles, experiences and practices of democratic organization. The children decide individually and in groups how, when, what, where and with whom to learn.	An educational project that makes contact with (external) Nature as its cornerstone principle. The discovery of the natural elements is the main stimulus to learn to do.	Public and private educational institutions that are addressed to the policies and programs of the MIUR.
Time management	Full freedom for children to follow their inclinations, preferences and curiosity.	Adaptation to family rhythms and the needs of minors. There are no pre-determined timescales. Often the learning topics are explored in depth until the child's curiosity is exhausted.	Free time management largely entrusted to the responsibility and accountability of the child who chooses what to do with his or her time	Great flexibility in the management of activities that are conceived and designed in a way that is appropriate to the age and composition of the pupils' group.	Detailed planning of activities (curriculum) and disciplines over a 10-month period of time that is not very flexible with regards to the class
The educational contents	Curiosity and personal talent are used as a driving force and springboard for a path of discovery and learning. The contents are agreed case by case.	Often ministerial programs are followed by maintaining a reference to teaching per year and per age group. Other times they proceed one subject at a time, dedicating months of study to a topic.	There are no predefined curricula; everything is agreed between educators and students. The content can be tailored to each pupil.	Flexible programming designed to respond to 4 fundamental categories of learning: curiosity, imagination, autonomy and creativity.	The learning of basic information and scientific information is aimed at the development of specific skills and social competencies.

The educational context	There is no predefined learning environment: you can learn in any context.	There is no predefined learning environment: you can learn in any context.	The learning environment is a building but with large open spaces available. Parents contribute directly to the furnishing and management of the chosen context.	Most of the activities take place outdoors. The building is only there for support.	Purpose built premises with classrooms for classes and activities planned according to different ages.
The relational context	Empathetic and unstructured; confidence in the child's instinctive ability to learn is communicated; initiative and cooperation between children and between adults and children is fostered.	Empathetic and unstructured confidence in the child's instinctive ability to learn is communicated; initiative and cooperation between children and between adults and children is fostered.	Focused on self-discipline, mutual respect, cultivating the sense of individual freedom.	Empathetic and unstructured; trust is communicated in the child's instinctive physical and relational/ cooperative abilities: entrepreneurship and cooperation between children and between adults and children is fostered.	Formal and structured according to roles and authority and therefore in the vertical sense. The horizontal relationships between equals often lead to the development of competitiveness and discriminatory situations between social and cultural differences.
The choice of method with regard to the role of the reference adult	The parent is called upon to respond to the learning needs of his/ her offspring by offering appropriate and welcome support to the child. The constitution of a network of families is useful to experiment and exchange educational strategies.	The parent-educator manages the child's education in a direct relationship without external distractions, according to the child's learning rhythms, paying attention to both the creative and the educational aspects.	The educator is a companion whose task is to support the child in a common process of investigation, discovery, knowledge and creative capacity.	The educator is a guardian whose responsibility is to help the child in a process of discovering his or her own abilities.	The teacher has a pivotal role also in learning process. Teacher's influence is determinant in choosing themes and activities. Teacher's evaluation is determinant in self-evaluation process.

The analytical observation of the homeschooling situation, and of other non-standard forms of education, was therefore achieved thanks to a sequence of two participating observations integrated with the analysis of the content of materials and texts: a substantial archive that aimed to raise awareness of the heuristic perspective of the researcher with regard to the reality of the investigation. The first participating observation session took place during the meeting in Rimini, S-Cool - Fourth Meeting of Parental Education, Rimini, 17-21 June 2016.

The second session was held at San Saba (ME) at the Second Meeting of Sicily Homeschooler on September 13-15, 2016.

In both sessions, the participating observation as a tool to collect empirical data about the investigated reality ensured (Gobo 2001: 18): a) the possibility for the researcher to establish a direct relationship with the social actors allowing opportunities to understand the meaning of the daily activities and the "non-standard" choice that he or she would otherwise not have had; b) that the individuals involved were observed without filters in their normal context wherever possible c) that knowledge was obtained without forcing their daily behaviors, d) the sharing not only of practices but also daily rituals, e) the learning of "codes" of such rituals, languages and behaviors that in addition to understanding the meaning of their actions allowed us to understand how they see and interpret the actions of "others".

Although the observation was the most important aspect of the information gathering phase, it was preceded and accompanied, as a matter of course, by the collection of different sources: in this case, websites and blogs on the homeschooling experience, Facebook pages, docu-films, interviews and short Youtube videos (diaries, letters, topics, organization documents, interviews, etc.).

With regard to the research in question, a field diary was compiled for both sessions of the participating observation relating to aspects considered relevant to the survey. The diary is a document in which were reported: a) some significant behaviors of the subjects observed; b) their way of approaching the training process; c) some conversations or comments considered to explain the decision to refuse children's inclusion in the state school; d) the description of the context in which the events took place and, sometimes, the description of the people involved because they are representative of a specific cultural affiliation.

The writing of the field diary played a flexible and rigorous role in the collection of qualitative data during observation in the field. The diary remains a faithful but creative record because it allows (even at a distance of time) a meditated and objective reflection on the activities and characteristics of the social units investigated.

In particular, the preference for participating observation rather than the "non-participating" option was in a certain sense obligatory: due to the particular nature of the phenomenon, involvement in daily activities and activities closely related to homeschooling (methods and materials, experiences and experiments in homeschooling training and teaching) was already planned because it was considered necessary for a concrete understanding of the phenomenon. For this reason, the participating research, rather than an investigation technique, could be defined as a way of being in the world (Gobo 2001: 15).

The description of the environmental context in which the two participating sessions took place and which opens the two sections of the field diary served to highlight the choices and the organizational methods of the homeschooler meetings and therefore contributed to outlining some cultural characteristics of the participants in the two different meetings. Furthermore, it should be pointed out here that the notes in the diary were made privately (usually late in the evening) out of discretion and to ensure credible and effective participation in the demanding demonstration and experimentation activities of the homeschooling "method" and lifestyle. Therefore, the diary does not report precise times but refers to the sequence of activities in which the researcher participated during the day, getting involved and actively contributing as part of the "homeschooling world".

> In order to note down the observations or graphic representations, I chose an elementary school notebook, which was quite anonymous though visible. During the observation, I realized almost immediately that it was not a good idea. We are newcomers to the world of homeschooling and its practices: among the participants we know only one family who was contacted electronically in the past months. Therefore, we are observed and listened to with particular attention and I sometimes sense mistrust. I am observed in the way I dress, the way I talk to the children, the way I sit on the floor. The notes, from now on, will be reported separately.
> (S-Cool - Fourth Meeting of Parental Education, Rimini, 17 June 2016).

> [...] I participated in two workshops: Lapbook Exhibition and Communicating Science. In both cases the perspective is turned upside down: it is the child who leads the learning path. In the first case, the workshop is held by an artist (a homeschooler mother for several years) of undoubted talent who, through examples and educational products, presents in a direct and clear way the method of learning by building the Lapbook on a theme or object of knowledge.

The method is simple and focuses on creativity and systematicity in finding and compiling information related to a chosen theme: this can be anything from bees to "how the motorcycle began". This method is particularly effective with children of ten years old and above who can express a certain aesthetic taste, numerous curiosities and thematic links by building the different parts of the Lapbook. The direction of the adult remains fundamental. I understand that the role of the director is different from that of a teacher, but not having experienced it directly I cannot evaluate the differences. Certainly, M.'s talent in this case makes the difference! All the works presented are captivating and full of information; the texts are excellent from an educational perspective and fascinating, full of colors, pockets, small windows and some with pop-ups. While presenting her works M. tells us about her decision and shares part of her family story. This creates an atmosphere of great collaboration, almost pioneering. [...]. This is what they say about themselves: the meeting itself actually opened with the statement "we are the pioneers".
(S-Cool - Fourth Meeting of Parental Education, Rimini, 18 June 2016).

The language in the participating observation is an important aspect of the observation experience and also one of the most difficult to manage properly. Its transposition into text for the field diary required some effort to adequately report the meaning of conversations containing common terms but with expressions that indicate the conflictual and oppositional relationship towards institutions and standard educational processes perceived as a limitation and a form of social and ideological control: «Metaphorically speaking, terms are like hooks that anchor a meaning and through them we can investigate the mental patterns and ways of reasoning of social actors». (Gobo 1999: 11)

> The password is... don't delegate. The family is the shield: homeschooling is for those who have decided not to delegate [the education of children], for those who have decided to take responsibility for their own lives and the lives of their children.
> (S-Cool - Fourth Meeting of Parental Education, Rimini, 18 June 2016)

During the sessions of scheduled meetings and activities, conversations with the participants were focused on the reasons for choosing unschooling or homeschooling and how they became aware of this opportunity to "manage" the educational path for their children. The organizational dimension was particularly discussed only with those individuals (mainly women) who clearly expressed their preference for homeschooling or for a different educational path from the one normally proposed in state schools (the standard path), such as the Montessori method. Where the rejection of the state school was clearer – perceived as a cultural imposition and control over the daily choices in lifestyle – rather than a conscious choice of the educational path that was the children's due, the conversation about the description of the daily values and habits that accompany this choice and the participation in the meeting(s) was initiated.

Some recurring themes arose from conversations during workshops and exchanges of confidence: a very critical cultural stance towards consumerist mass culture which often goes hand in hand with a political opinion close to anarchist positions; a cluster of freelance professionals close to the world of arts and crafts (musicians, street artists, buskers, freelancers, and permaculture practitioners); an alternative (and sometimes syncretic) kind of spirituality that strongly favors a holistic vision – interconnected with the elements of nature – (Hanegraaff 2005; Tucker 2002) that often, in a counter-cultural perspective, rejects traditional moral values by sanctifying the individual-nature relationship (Houtman and Aupers 2007: 305-310); and a lifestyle based on unconventional consumption choices (home farming, permaculture, vegetarianism, veganism, barter, responsible consumption, and recycling). However, in all perspectives the state school is considered the necessary counterpart from which to escape and set oneself apart.

The family is considered to be the founding element of the daily life experience that enables the construction of a group identity based on difference. It represents the antidote to the criticalities embodied by the public school system, structured and formal, insensitive to the uniqueness of the individual: this belief is often accompanied by the failed or traumatic school memories of homeschooler parents.

The two subsequent participating observations also led to a change in the main nature of the field diary. If during the first meeting (Rimini) the copious observational notes were justified by the need to describe basic social practices, micro-events and daily micro-actions (Gobo 1999: 7), the participation in the second Homeschooling Meeting in Sicily required more attention to theoretical notes and notes of an emotional nature (Gobo 1999: 8-10).

> In the end, it all depends on what your lifestyle choice is...
> First of all, we had to get rid of the stuff... of normality...
> (Second Meeting of Sicily Homeschooler Families, San Saba, 15 September 2016)

The formulation of these statements led to further and deeper considerations with regard to the evident dynamics of organization and management of the educational process, focusing on the role that is highlighted within these daily dynamics more than others: the role of the mother.

17.3 *Us and the Others: the leading role of mothers*

Starting from a database approach (Strati 1997), the dense set of ethnographic notes was used to create a synthesis of the elements and typological contents (Morton 2010) of the different manifestations of non-standard training courses. The simultaneous and cross-referenced reading of the notes highlighted characteristics that recur in several cases with regard to the distinctive elements of the different educational models (Leonora 2019: 137) and this suggested a further typological formulation of the different approaches starting from the maternal perspective (De Sanctis, Fariello and Strazzeri 2020). The educational decision of the homeschooler mothers we met is rooted in the dynamics of social reproduction if we consider that "microhistory is the raw material of the great story" (Ferrarotti 1999: 11). This decision is a precious indicator of what constitutes a determining factor of their relationship with the wider society: the family (even a single parent!), as a composite social unit able to generate relational processes and social actions through relational dynamics and material resources, offers a possible path to the construction of social identity (Sciolla 2010).

As stated by all the mothers we met during the participating observations, homeschooling practices are a "challenge to the system". The interest in this case focuses on the particular conflictual dynamic that takes place between two fundamental educational institutions: the school and the family. Indeed, in its clear intention of detachment from the educational institution, the homeschooler family distinguishes itself as a hyper-functional family (Leonora 2014) because it is concentrated on the transmission of cultural representations (life choices) and knowledge instrumental to its own uncontaminated presence in the social structure from which it can stand out and separate itself.

> This morning the Foresta[2] family (4 pre-adolescent children) told their story and explained their homeschooler choice. [...] Starting from the decision to move to the countryside, everyday family life becomes the cornerstone for planning and organizing resources, relationships and practices of their life project and the cohesive and close-knit "family group" represents the point where the life dimensions of its members merge. They are six very different but tightly knit entities. They have different interests that require different commitments: music (on a professional level), dance, sport, social work (as a

---

2   Invented name.

doctor!). However alternative they may be, they are convincing: the maturity of the teenage girls is particularly persuasive
(Second Meeting of Sicily Homeschooler Families, San Saba, 16 September 2016).

The main purpose of homeschooling is not education, but awareness parenting (Arace and Scarzello 2010): a high awareness of the duty of care, an educational model that, unlike the traditional approach, recognizes the uniqueness of the individual and his primary need to realize himself in a social environment that most of all safeguards his freedom of choice and cultural expression. One of the fundamentals of homeschooling is respect for the natural curiosity of the child who should follow and nurture his or her talents and inclinations through knowledge appropriate to his or her learning rhythms and abilities. Unfortunately, when a child enters state school, this does not happen (Robinson 2001), because times, subjects and notions are imposed, perhaps in that phase of personal growth in which the child is not ready (Di Martino, 2017).

The deep and continuous bond that homeschooling mothers establish with their offspring operates an osmotic continuity in the two formally distinct processes of socialization and education, nourishing an ambivalent reflective potential: on the one hand a critical thought; on the other hand a self-referentiality consistent with the state of a hyper-functional family.

Ethnographic data show that the main problem that concerns non-standardized mothers is the controversial approach to normality that surrounds the homeschooler family that is required to achieve a contiguous and parallel path while maintaining its small group identity (Sciolla 2010). In this case, the educational path and the process of socialization overlap continuously because they are substantially characterized as learning and internal appropriation of meanings and more general rules that define a particular *weltanschauung* – the perception of the world as a meaningful reality.

The Promethean myth of ineluctable progress at any cost – thanks to the knowledge bestowed by the system – is courageously challenged by these "goddesses" who direct their material and relational resources to trace a path of opportunity and protection for their offspring in the dangerous and contradictory world of humans (see summary table).

The analytical categories inspired by these mythological figures are taken directly from the conversations held during the two participating observations (§ 2). Each type/character tries to summarize an ideal approach that is formed by an operative choice between homeschooling and unschooling.

Athena summarizes the mother who expresses primarily active protection towards her offspring and the family: she is a practical organizer who does not delegate; rather, she chooses and uses in a strategic way the aspects of everyday life to achieve her educational goals. Often her family is bilingual (perhaps thanks to a mixed marriage) and she shows an open mind and an expert use of technology and social media. The impossibility of obtaining a response to the demands for flexibility and adaptability from the standard educational model (especially the state model) is decisive in the choice to practice homeschooling. The approach to knowledge remains rational and she often uses the same (digital) information channels shared by public educational institutions, encouraging the use of the latest technology also for innovative and productive purposes. Although in Italy one cannot speak of a "homeschooling movement" (Collom and Mitchell 2005) Athena considers the contact and agreement with the other homeschooler families present in the territory, connected by periodic events of comparison and mutual support, as instrumental to the recognition of its specificity.

> I can't delegate to others because it's something only I can do...
> The child is curious for sure, but often not with the timing of the school curriculum.
> (S-Cool - Fourth Meeting of Parental Education, Rimini, 2016).

The Aphrodite mother represents the arrival point of a radical choice that starts well before and beyond the experience of motherhood: what characterizes her is a holistic perspective, not mainstream, and often oriented by a syncretic spiritualism close to Eastern culture and spirituality. Her cultural orientation is expressed in alternative choices of life and consumption, in the counter-cultural sense, which for instrumental reasons are adapted to the social and geographical context of reference. The radical diversity of this lifestyle is well coordinated with the ideals and ways of managing homeschooling and even more so unschooling (Holt 1976). Her relationship with her offspring is based on empathic, emphatic, and emotional dimensions. Primary socialization is never interrupted and an instrumental formative path is inserted in it in order to guarantee first of all a position of independence of the individuals with regard to society; the knowledge and the training path to reach it may be customized and limited to the talent and the specific interests of the family and the offspring (as in the case of artists, performers and inventors).

> The conversation at the mother-child bond discussion group was always light and surreal, charged with empathic emphasis. Many mothers of many

different ages shared their experience of pregnancy and childbirth. [...] I had a conversation with a young mother in her thirties who was about to embark on this experience of unschooling with her eldest son (5 years old) and in the momentary intimacy of the situation I said that her face reminded me of someone; in fact, to tell the truth, I was convinced I knew her. But this being impossible, I added that perhaps we had met in another life. My joke was taken seriously, literally. Our level of confidence increased a lot and we completed the work of aesthetic composition together: we made a mum-mandala (a scrapbook of pieces of newspaper on white paper). I gave it to her, but I wish I hadn't. I should have taken a photo of it and kept it. It was colorful and representative of an indefinite spirituality but respectful of all forms of life, open to knowledge as an encounter. The evocative image helped to talk more about her experience of motherhood and the period of pregnancy during which she was lucky enough to be joined by a Doula. For me this was the first time I had heard about the figure and the role of the Doula. [...] In the special relationship that developed between them, the Doula recommended the un-schooling experience as the best management of two different learning ages (the two children).
(S-Cool - Fourth Meeting of Parental Education, Rimini, 19 June 2016).

Demeter is the most complete representation of the homeschooler mother. Demeter represents the mother bearer of the only original bond of care that is the true pivot of the family organization: she is the center and the undisputed engine of the family. Her charismatic centrality can have different roots, ranging from radical religious choices (which do not like culturally diverse environments), to cultural and ideological choices often of an ecological and naturalistic nature, but expressed with logical and scientific consistency (all the subjects who have inspired this figure promote a style of responsible eating and consumption). According to the radicality that distinguishes this figure, normal attendance at state schools is considered a coercion or limitation that exposes the offspring to a cultural model that is not shared by the families and not considered useful to ensure a dignified future for the offspring. Her resistance and opposition to mass education is motivated by a desire for change and improvement following timescales and aims that are different from those of the dominant culture (Bell hooks 1994): Demeter shows a feminist consciousness, and she is aware of the traps inherent in universal and standardizing education. However, she recognizes the necessary contribution of knowledge for the affirmation of individuality, so she chooses and creates the best teaching model for her offspring by following her child's unique characteristics. For this reason, Demeter is always willing to experiment with effective communication channels towards all those involved (even the state school) and does not exclude proximity with them, as long as mutual respect is guaranteed. Unlike the other types, Demeter is interested in promoting her own teaching model not for proselytism but for divulgation purposes, to show and share her own awareness about the most innate and effective way of experiencing family life and the educational process.

> We moved [to a cottage with a garden, ed.] in the country because this is the best way to live and grow up.
> My son has special needs. The contact with him allows me to understand him: I already know when the time is right...
> (Second Meeting of Sicily Homeschooler Families, San Saba, 2016).

In the overview of these typological figures, the homeschooler mother, as a social phenomenon, offers considerable insight into the internal dynamics of the educational path created independently for her offspring, as opposed to the standard educational program. In the alternative and unconventional practice of the educational process organized outside the institutional context of state schools, the design function of an entity that organizes resources, relationships and practices of daily life in which the cultural and normative dimensions are foundational and characterizing (Saraceno 2017) is highlighted. However, according to this particular heuristic reading, a central, powerful and ambivalent female subjectivity is defined (Pulcini 2001: 109 et seq.). The centrality of the maternal figure is constantly confirmed by her ability to understand and manage the needs linked to the maturation, and therefore education, of her offspring. This capacity is expressed in different ways that in actual fact can manifest more or less strong variations of the three types proposed here. The more the differences are intensified, the more the chosen educational path offers not only training but also identity. In recognizing the bond with her offspring, the homeschooler mother affirms her independence from the dominant society and culture: she does not cut her own narcissistic mother-child cord that pours out her own aspirations on her offspring, but definitely cuts the narcissistic cord of the individual with the society that wants him/her to be a mere micro-reproduction of itself (McLuhan 1951: 247).This is the soft revolution of the homeschooler mothers: not a child to admire and to be admired (McLuhann 1951: 155) but a child to be freed, and to whom the educational path offered is aimed primarily at the defense of individuality and individual well-being in the face of the depersonalization of the institutionalized mass.

Table 2 - Types of homeschooler mothers

Ideal representation Relationship with offspring	Athena	Aphrodite	Demeter
Reference principle	Responsibility	Affection-emotivity	Caring relationship
Habitat	Instrumental: equipped according to educational objectives.	Expressive: chosen or prepared to nurture emotions or sensations deemed positive.	Participative: organized so as to involve individuals and tools in the educational process.
Choice of teaching model	The parent enhances the specificities of the biographical period (childhood, adolescence, puberty) and responds to the learning needs of the offspring by offering information, direct experiences and opportunities for experimentation appreciated by the children. The establishment of a network of families is useful for experimenting and exchanging educational opportunities and strategies.	The parent-educator manages the child's education in a direct relationship without external distractions, according to the child's learning rhythms, taking care of both the creative and the educational aspects.	The educator is a companion who has the task of accompanying the child in a shared process of investigation, discovery, and knowledge, and underlines the role of the family as cultural entity in socialization and inside the educational processes.

Bibliographical references

Arace, A., and D. Scarzello, 'Maternal Parenting Practices in School Age. A Study About Social And Didacting Caregiving', *International Journal* of *Developmental and Educational Psychology, INFAD Revista de Psicología*, 1 (2010), 149-158.

Bell hooks, *Teaching to transgress: education as the practice of freedom* (New York: Routledge, 1994).

Besozzi, E., *Educazione e società* (Roma: Carocci, 2017).

Bottani, N., *Requiem per la scuola? Ripensare il futuro dell'istruzione* (Bologna: Il Mulino, 2013).

Charmaz, K., *Constructing grounded theory* (London: Thousand Oaks, Sage Publications, 2006).

Chistolini, S., 'Apart from the Steiner School and Montessori Method. Homeschooling is the answer for families to the social crisis of schools', *New Jersey Journal of Supervision and Curriculum Development*, 53 (2009), 46-56.

Codello, F., and I. Stella, *Liberi di imparare. Le esperienze di scuola non autoritaria in Italia e all'estero raccontati dai protagonisti* (Firenze: Terra Nuova, 2011).

C O E, *White Paper On Education And Training, Teaching And Learning Towards The Learning Society* (Bruxelles: 1995).

Collom, E., and D. E. Mitchell, 'Home schooling as a social movement: Identifying the determinants of homeschoolers' perceptions', *Spectrum*, 25 (3) (2005), 273-305.

Colombo, M., *Gli insegnanti in Italia. Radiografia di una professione* (Milano: Vita e pensiero, 2017).

Delors, J., *Learning: the treasure within* (Paris: Unesco, 1996).

De Sanctis, D., Fariello, S., and I. Strazzeri, *Sociologia della maternità* (Milano: MIMESIS, 2020).

Di Martino, E., *Homeschooling. L'educazione parentale in Italia* (CreateSpace Independent Publishing Platform, 2017).

Di Pol, R. S., *La scuola per tutti gli italiani. L'istruzione di base tra Stato e società dal primo Ottocento ad oggi* (Milano: Mondadori, 2016).

Ferrarotti, F., *La verità? È altrove* (Roma: Donzelli, 1999).

Gaither, M., 'Homeschooling in the USA: Past, Present, and Future', *Theory and Research in Education*, 7 (3) (2009), 331-346.

Gaither, R. M., *Homeschool. An American History* (New York: Palgrave Mc Millan, 2017).

Gamuzza, A., 'Homeschooling in Italy. Evidences and Challenges from the Field,' *Supplemento a Diritto & Religioni*, 2 (8) (2013), 3-11.

Glaser, B., and A. Strauss, *La scoperta della Grounded Theory. Strategie per la ricerca qualitativa*, ed. by A. Strati (Roma: Armando, 2009).

IEP, *Global Peace Index. Measuring Peace in a Complex World* (Sydney, 2020).
Gobo, G., 'Le note etnografiche: raccolta e analisi', *Quaderni di sociologia*, 21 (1999), 1-21.
— *Descrivere il mondo. Teoria e pratica del metodo etnografico* (Roma: Carocci, 2001).
Gobo, G., and A. L. Tota, *L'osservazione partecipante. Un metodo per le scienze sociali* (Roma: Carocci, 2000).
Hanegraaff, W. J., 'New age', in L. Jones (ed.), *Gale encyclopedia of religion* (Chicago: McMillan, 2005), pp. 6495–6500.
Holt, J., *Escape from Childhood* (Boston: Dutton, 1974).
Houtman, D., and S. Aupers, 'The spiritual turn and the decline of tradition. The spread of post Christian spirituality in 14 Western countries, 1981–2000', *Journal for the Scientific Study of Religion*, 46 (3) (2007), 305-320.
Jolly, J. L., Matthews, M. S., and J. Nester, 'Homeschooling the Gifted A Parent's Perspective', *Gifted Child Quarterly*, 57 (2) (2013), 121-134.
INVALSI, 'Rapporto Prove INVALSI 2019', in: https://invalsiareaprove.cineca.it/docs/2019/Rapporto_prove_INVALSI_2019.pdf
ISTAT, 'Rapporto SDGs 2019. Informazioni statistiche per l'Agenda 2030 in Italia', in: https://www.istat.it/it/files/2019/04/SDGs_2019.pdf
Kunzman, R., and M. Gaither, 'Homeschooling: A Comprehensive Survey of the Research, Other Education', *The Journal of Educational Alternatives*, 2 (2013), 4-59.
Leonora, A. M., 'Mamma ho capito come funziona la scuola. Posso restare a casa. Per una prospettiva sociologica sulle pratiche homeschooling in Italia', *Annali della facoltà di Scienze della formazione, Università degli studi di Catania*, 13 (2014), 61-78.
— 'Veder crescere un albero. la prospettiva capovolta dei processi educativi non standard in Italia', *Annali della facoltà di Scienze della formazione, Università degli studi di Catania*, 18 (2019), 127-139.
McLuhan, M., *La sposa meccanica. Il folklore dell'uomo industriale* (Perugia: Sugarco, 1951, repr. 1986).
Morton, R., 'Home Education: Constructions of Choice', *International Electronic Journal of Elementary Education*, 3 (2010), 45-56.
Murphy, J., *Homeschooling in America. Capturing and Assessing the Movement* (Thousand Oaks: Corwing, 2012).
OECD, 'Education at a Glance. Uno Sguardo sull'Istruzione (Italia)', *OECD Indicators*, (Paris: OECD Publishing, 2019), https://doi.org/10.1787/f8d7880d-en.

Pulcini, E., *L'individuo senza passioni. Individualismo moderno e perdita del legame sociale* (Torino: Bollati Boringhieri, 2001).

Robinson, K., *Out of Our Minds. Learning to be Creative* (Capstone Ltd., 2001).

Saraceno, C., *L'equivoco della famiglia* (Roma-Bari: Laterza, 2016).

Sciolla, L., *L'identità a più dimensioni. Il soggetto e la trasformazione dei legami sociali* (Roma: Ediesse, 2010).

Silverman, D., *Doing Qualitative Research. A Practical Handbook* (London: Sage, 2000).

Snyder, M., 'An Evaluative Study of the Academic Achievement of Home-schooled Students Versus Traditionally Schooled Students Attending a Catholic University', *Catholic Education: A Journal of Inquiry & Practice*, 16 (2) (2013), 288-308.

Strati, A., 'La «Grounded Theory»', in L. Ricolfi (ed.), *La ricerca qualitativa* (Roma: La Nuova Italia Scientifica, 1997), pp. 125-163.

Tucker, J., 'New age religion and the cult of the self', *Society*, 39 (2) (2002), 46–51.

Valero, E. M. J., 'The Long Way Home: Recent Developments in the Spanish Case Law on Home Education', *Oxford Journal of Law and Religion*, 3 (1) (2013), pp. 127-151.

Wax, R. H., *Doing Fieldwork* (Chicago: University Press, 1971).

Authors Biographical Profiles

Tova Benski is a senior lecturer emerita and the Emerita Dean of the School of Behavioral Sciences at the College of Management – Academic Studies, Rishon Leziyon Israel. Her fields of academic interest and research include gender, social movements, peace studies, and the sociology of emotions. She is the co-author of the book internet and emotions (Routledge 2013), and co-editor of Current Sociology special issue (2013. She served twice as the president of RC 48 (2002-2007 and 2014-2018). Currently she is an elected member of the executive council (EC) of the ISA and member of the Board of RC 48, RC 36 and TG08 of the ISA.

Rita Bichi is full professor at the Università Cattolica del Sacro Cuore di Milano (UCSC), School of Political and social sciences. She belongs to a scientific area SPS/07 (General Sociology). She teaches Epistemology, Methodology of social research and Sociology in the Schools of Sociology and Political Science. She has conducted research nationally and internationally, participating, as director, in projects funded by the European Union and by Italian and foreign institutions and organizations. Bichi has authored numerous books, book chapters and articles on Italian and international journals.

Angela Carbone is Associate Professor of Modern History at the Department of Education, Psychology, Communication of the University of Bari Aldo Moro. Her main areas of research are: family; gender history; charity and assistance. Among her list of publications are: *Vita nei Sassi. Famiglia, infanzia e assistenza a Matera in età moderna* (Cacucci Editore, 2005); *Ritirate dalle cose del mondo. Donne e istituzioni nel Mezzogiorno moderno* (Guida Editori, 2020).

Giada Cascino has a PhD in Social inclusion in multicultural contexts. Lecturer in the subject for the SSD Sector SPS/07 – General Sociology at the University of Enna "Kore". Member of national and international research groups: her study and research interests focus on the role of institutions in the process of social integration in response to the challenges posed by social change.

VINCENZO CICCHELLI is an Associate Professor at University Paris Descartes and Research Fellow at Centre Population et Développement (CEPED) (Université de Paris / Institut de Recherche pour le Développement (IRD)). He is currently the Director of International Relations at GRIP (Global Research Institute of Paris, University of Paris). At Brill, he is the Editor-in-Chief (with Sylvie Octobre) of the 'Global Youth Studies' suite: http://www2.brill.com/gys. He is the author of many books and articles, of which the latest are (with Sylvie Octobre and Viviane Riegel, eds.) *Aesthetic Cosmopolitanism and Global Culture* (Brill, 2019).

MARIAELENA COSTA is PhD in Modern History. She cooperates with full professor of Modern History, Department of Education, University of Catania. Her interests are about urban history and economic structures. She has several scientific publications to her credit. Among her publications: Il *Liber Gratiarum et Privilegiorum* dell'*Universitas* di Nicosia, in E. Frasca (a cura di), *Il valore e la virtù. Studi in onore di Silvana Raffaele*, Acireale-Roma, Bonanno, 2019.

GIOVANNA DA MOLIN is Professor Emeritus of Modern History at the University of Bari Aldo Moro. Her research interests focus on: population; family; foundlings and orphans. In particular, the following publications should be mentioned: *I figli della Madonna. Gli esposti all'Annunziata di Napoli. Secc. XVII-XIX* (Cacucci Editore, 2001); *Famiglia e infanzia nella società del passato. Secc. XVIII-XIX* (Cacucci Editore, 2008); *Storia sociale dell'Italia moderna* (Editrice La Scuola, 2014).

LIANA M. DAHER is Full Professor of Sociology, Department of Education, University of Catania. Her main research fields are collective action and social movements, migration and multicultural citizenship. She is President of the RC48 (Social Movement, collective action and social change) of the ISA. She has authored books, book chapters and articles on Italian and international journals. She has taken part, both as principal investigator and member, to several national and European research projects. Currently, she covers the role of scientific coordinator for the Erasmus+ project KA3 Support for policy reform "TIEREF - Toward inclusive Education for Refugees children and the

H2020 "PARTICIPATION. Analyzing and Preventing Extremism Via Participation."

Paolo De Nardis is full professor of sociology at the Sapienza University of Rome where he was also director of the Department of Sociology and dean of the faculty of the same name. He is the doyen of the Italian Academic Sociologists and President of the Institute for Political Studies, S. Pio V in Rome. His research interests range from the analysis of the foundations of social sciences to history and sociological theory, to the analysis of complex structures, to the issue of conflict and new urban science. He is the author of over two hundred publications.

Antimo Luigi Farro is Full Professor at Sapienza University of Rome, Italy, Department of Social Sciences and Economics (DISSE). He teaches urban sociology and is currently a Faculty Socrates Scientific Coordinator, and member of CADIS International. He published several books and articles related to social conflicts and social movements, urban issues, local communities, and immigration. He just published (2020) "Il mondo in un quartiere. Migrazioni internazionali Esquilino Roma", with CEDAM/Kluwer.

Elena Frasca is researcher in Modern History at the University of Catania. She studied institutions, medicine, relations between universities and urban power, cultural and scientific academies, *Grand Tour*, criminal legislation, gender history. She has published many articles, essays and monographs on subjects of her studies. For the monograph *L'eco di Brown, Teorie mediche e prassi politiche*, Roma, Carocci, 2014, she won the award "Elide Stramezzi" (2015) assigned by the "Accademia di Storia dell'Arte Sanitaria", in Rome.

Augusto Gamuzza is Assistant Professor (with tenure track) in Sociology at the University of Catania, Department of Education and researcher at ISIG (Institute of International Sociology, Gorizia) and CUrE (Interdepartmental Research Center for the Community University Engagement- University of Catania). He is the Scientific director of the "OfficinaSociale" research laboratory and board member of the NGO CO.P.E. (Cooperation in Emerging Countries), Catania since 2016.

Gennaro Iorio, Ph.D., is Full Professor and Head of Department of Political and Social Studies, at University of Salerno (I). He teaches History of Sociology and Sociology at bachelor, Sociology of Innovation at masters courses and Sociology at Ph.D. Courses. Member of the board of Ph.D. Digital Innovation, Social Theory and Public Policy. His research interests are Social Theory and Critical Theory, Digital Innovation and Poverty. He has published over fifty scientific paper and books.

ANNA MARIA LEONORA is researcher in Sociology at University of Catania, Department of Education. Her main research interests are: solidarity and socialization dynamics; alternative forms of educational processes – with a special focus on homeschooling practices and communities in Italy; the link between social changes and educational practices in postmodern society applying ethnographic and mixed method approaches. Since 2010, she works as senior researcher for the Erasmus + projects as NORADICA - Inter-Religious Dialogue Against Radicalization of Youth through Innovative Learning Practices at School.

SIMONE MADDANU earned his PhD at School for Advanced Studies in Social Sciences (EHESS) of Paris, France, in 2009. He teaches sociology and contemporary social problems at University of South Florida, Tampa, and he is a member of CADIS International. He recently published two monographs and several peer reviewed articles on social movements, immigration, Islam in Europe, urban studies, populism, and modernity.

GIORGIA MAVICA is a Research Fellow in General Sociology SPS/07, PhD in Human Sciences and Expert in General Sociology at the Department of Education, University of Catania. Her current research interests focus on the theme of youth subjectivity expressed through collective action and the consequences of such action. She collaborates, and has collaborated, as Junior Researcher in the projects: TIEREF (KA3 Erasmus +), Unaccompanied and Separated Children in their Transition to Adulthood in Italy (UNICEF), NORADICA and Multicultural Schools (KA2 Erasmus +). She has several scientific publications to his credit.

DAVIDE NICOLOSI is a PhD student in Educational processes, theoretical-transformative models and research methods applied to the territory at the Department of Education Catania. His fields of interest concern the collective action and the protests, in particular the social movements and the associations of second generations of migrants. He collaborates, and has collaborated, as Junior Researcher in the projects: TIEREF (KA3 Erasmus +), Unaccompanied and Separated Children in their Transition to Adulthood in Italy (UNICEF), NORADICA (KA2 Erasmus +).

SYLVIE OCTOBRE is a researcher at Département des études, de la prospective et des statistiques, French Ministry of Culture, and Research Fellow at GEMASS/CNRS. At Brill, she is the Editor-in-Chief (with Vincenzo Cicchelli) of the 'Global Youth Studies' suite : http://www2.brill.com/gys. She is the author of many articles and books, of which the latest are (with Vincenzo Cicchelli and Viviane Riegel, eds.) *Aesthetic Cosmopolitanism and Global Culture* (Brill, 2019).

SILVANA RAFFAELE is full professor at the University of Catania. Specialist in the Bourbon period, she has studied subjects about Southern Italy social, political and institutional history, in particular about 18th and 19th centuries' laws, demography, family's structures, medical and university teaching linked with urban political life, education, forced nuns. Some of her publications: *Famiglie e senza famiglia...* (Napoli, Esi, 2000); *La bottega dei saperi...* (Acireale-Roma, Bonanno, 2005); *Aut virum... aut murum...* (Acireale-Roma, Bonanno, 2010).

CINZIA RECCA is Lecturer in Early Modern History at the University of Catania in the Department of Education. Her main field of research include the European Enlightenment specially regarding feminine roles. Among her recent publications: "The reversal of dynasties during the Bourbon era in the Kingdom of Naples." in A.M. Rodrigues, M. Santos Silva, J. Spangler (eds). Dynastic change Legitimacy and Gender in Medieval and Early Modern Monarchy. London – New York: Routledge, (2020).

ALESSANDRA SCIERI is currently free worker at the Department of Educational Sciences, University of Catania. Her current research interests focus on the theme of ethics of responsibility in young people and environmental issues. She collaborates, and has collaborated, as Junior Researcher in the projects: TIEREF (KA3 Erasmus +), Unaccompanied and Separated Children in their Transition to Adulthood in Italy (UNICEF), NORADICA (KA2 Erasmus +).

SERGIO SEVERINO is Full Professor of General Sociology and Coordinator of the Course of Studies in Social Service and Criminology at the University of Enna "Kore". In the same University, he held the position of Coordinator of the XXV cycle PhD in Sociology of Innovation and Development. He has also taught at Italian and foreign universities (Messina, Palermo, Catania, Riga, Warsaw, Budapest, Dobrich, Braganza, and Siauliai).

GIOVANNA SUMMERFIELD is professor of Italian and French and Associate Dean for Educational Affairs in the College of Liberal Arts at Auburn University. Her research interests include eighteenth-century French and Italian literature, Mediterranean history and culture, film studies and women's studies. Recent publications are: *Sicily on Screen. Essays on the Representation of the Island and Its Culture* (2020) and *Sicily and the Mediterranean: Migration, Exchange, Reinvention* (2015).

MIMESIS GROUP
www.mimesis-group.com

MIMESIS INTERNATIONAL
www.mimesisinternational.com
info@mimesisinternational.com

MIMESIS EDIZIONI
www.mimesisedizioni.it
mimesis@mimesisedizioni.it

ÉDITIONS MIMÉSIS
www.editionsmimesis.fr
info@editionsmimesis.fr

MIMESIS COMMUNICATION
www.mim-c.net

MIMESIS EU
www.mim-eu.com

Printed by
Digital Team – Fano (PU)
November 2020